SAP Excellence

Series editors:

Prof. Dr. Dr. h.c. Peter Mertens,
University of Erlangen-Nürnberg

Dr. Peter Zencke,
SAP AG, Walldorf

Springer
Berlin
Heidelberg
New York
Barcelona
Hong Kong
London
Milan
Paris
Singapore
Tokyo

Jörg Becker · Wolfgang Uhr
Oliver Vering

Retail Information Systems Based on SAP Products

With Contributions by
Lars Ehlers, Ernest Kosilek, Matthias Lohse
and Stefan Neumann

With 105 Figures
and 7 Tables

Springer

Prof. Dr. Jörg Becker
University of Muenster
Department of Information Systems
Steinfurter Straße 109
48149 Muenster
Germany

Prof. Dr. Wolfgang Uhr
Dresden University of Technology
Department of Economics and Business Administration
Mommsenstraße 13
01062 Dresden
Germany

Dipl.-Wirt.-Inform. Oliver Vering
University of Muenster
Department of Information Systems
Steinfurter Straße 109
48149 Muenster
Germany

ISBN 3-540-67199-4 Springer-Verlag Berlin Heidelberg New York

Library of Congress Cataloging-in-Publication Data applied for
Die Deutsche Bibliothek – CIP-Einheitsaufnahme
Becker, Jörg: Retail Information Systems Based on SAP Products: With 7 tables / Jörg Becker; Wolfgang Uhr; Oliver Vering. With Contributions by Lars Ehlers ... – Berlin; Heidelberg; New York; Barcelona; Hong Kong; London; Milan; Paris; Singapore; Tokyo: Springer, 2001
(SAP Excellence)
ISBN 3-540-67199-4

Springer-Verlag Berlin Heidelberg New York
a member of BertelsmannSpringer Science+Business Media GmbH
http:/www.springer.de

Hardcover-Design: Erich Kirchner, Heidelberg

SPIN 10762662 42/2202-5 4 3 2 1 0 – Printed on acid-free paper

Preface

Times for retailers have become difficult. Stagnating real incomes of the consumers have resulted in stagnating sales for retailing companies and so produced a massive competitive pressure. Concentration and market power of the large companies determine the developments in retailing. Changed values, rising price awareness by consumers, rapidly decreasing customer loyalty, internationalization of the procurement markets, increasing cost pressure with the need to reduce margins, structural changes to more efficient company forms – the list can be continued without end: retailing companies in their role as intermediary between producers and consumers nowadays face challenges as never before. They react with changed business concepts and organizational solutions, cross-company cooperation and – as essential prerequisite – with the use of modern information and communications technologies.

No other technical development has affected and changed retailing more than computer-based information and communications systems. They are not only essential in a dynamic retailing world for the rational acquisition, control and monitoring of the goods flows. Not at least through the increased data volumes resulting from scanning in merchandise management systems, a valuable raw material for obtaining information is provided for the marketing support and retail controlling. New business concepts such as Efficient Consumer Response, Supply Chain Management or Category Management require the electronic exchange of data and documents and thus inter-company linked information systems that support the business processes along the value chain. In companies with branches, the increased customer orientation, flexibility and ability to respond have resulted in a trend to more decentralization of competencies. Central coordination must be combined with decentral processing and with local or regional oriented micro and segment marketing. "Electronic commerce" is gaining in importance: the Internet will not replace the traditional purchasing channels but will supplement them and thus also provide a partial substitute.

All the developments mentioned here demand innovative integrated application systems that embody the current state of theoretical knowledge and the best business practices from the industry.

As result of the historic development of data processing in many retail companies, heterogeneous data processing systems still often co-exist with various subsystems. However, "home grown systems" and specialized dedicated solutions have a major disadvantage: they are not easy to integrate, the exchange of information between these applications is difficult, and even the user interfaces are different. Integrated business standard software with defined interfaces for the connection of "external systems" such as POS systems, systems for warehouse optimization or online stores, that can be customized to the special requirements of the various retail companies can provide help here. We can also observe today in the retailing world the further accelerating trend in the use of standard software. Such software allows the internal integration within the retail companies and the external integration with partners. In addition to industry-neutral software, industry solutions are increasingly being offered. Retail information systems support retailing-specific processes such as purchasing, logistics, supply, sales and promotions, in addition to the managerial information tailored to retailing.

The large ERP solutions from the industry environment could not cover, or only inadequately, the specific requirements of retailing. SAP also initially underestimated these problems. The original schedule for the development of a dedicated retailing solution based on the SAP R/3 standard solution could not be held. In addition, the first Retail version (Retail 1.2 B) provided inadequate support for many retailing-specific requirements. This meant that the first pilot customers were confronted with complicated procedures, an almost insurmountable data updating effort and performance problems.

In the meantime, SAP, with extensive personnel and financial effort, has solved many of these problems and greatly extended the retailing-specific functionality. The current version, SAP Retail 4.6, is increasingly establishing itself as being the leading solution for mid-sized and large retail companies. This means it can be rightly described as being an integrated *retail* information system.

With this book we have tried to present in compact form the important concepts and functions of SAP Retail based on the representation of the basic requirements made on modern integrated retail information systems and taking account of current developments such as Efficient Consumer Response (ECR) and Category Management (CM). This should show the functions and capabilities offered by integrated retail information systems in general, and by SAP Retail in particular.

Although, this is already apparent, this book is characterized by a heavy SAP orientation, it is not an SAP Retail user's manual. Our central idea is to structure the major retail-specific details and to describe these from the business viewpoint. We then show those readers that are considering, or have already started, the use of SAP Retail how SAP Retail represents these details. In this way we show how modern business concepts for retailing companies can be realized using innovative standard software.

Chapter 1 briefly describes the current market and competitive conditions in retailing. Chapter 2 presents with the retail H-model a general framework for the structuring of integrated retail information systems. This framework is used as basis in Chapter 3 to consider the retailing-specific basic functionality of SAP Retail. The following chapter presents concepts for the realization of distributed merchandise management systems on the basis of SAP Retail. This is a question of central importance for retailing companies and retailing concerns with branches. Chapter 5 provides a detailed discussion how SAP Retail can be used to obtain information to support managers. It discusses both the topics of the retail management system (RIS) and also the new SAP Business Information Warehouse (BW). The final two chapters present current trends and developments in the area of electronic commerce, and for the planning and control of goods flows between industry and retailers. They also show how this can be realized on the basis of SAP Retail. This book appears as part of the SAP Excellence series, in which the following books have already been published:

- Appelrath, Hans-Jürgen; Ritter, Jörg: SAP R/3 Implementation. Methods and Tools,
- Knolmayer, Gerhard; Mertens, Peter: Supply Chain Management Based on SAP Systems. Order Processing in Manufacturing Companies,
- Buxmann, Peter; König, Wolfgang: Inter-Organizational Cooperation with SAP Systems. Perspectives on Logistics and Service Management

A feature of all these books, and thus also this book, is that employees of SAP have recently validated all statements made about the software. This not only permitted the authors to take account of the current developments of SAP products, but also meant that information was provided for many detailed questions that normally would not have been available to authors who are not employees of SAP.

At SAP, we would particularly like to thank Uwe Gerlach as central contact partner who actively supported us in the checking and updating of the manuscript. From the co-workers at our institutes, we would like to mention Lars Ehlers and Stefan Neumann in Münster, and Ernest Kosilek and Matthias Lohse in Dresden. They enthusiastically provided and prepared information, and were involved in the formulation of some sections. Marcus Kunigk helped us with the layout and the preparation of the graphics. We would like to express our gratitude to you all and our conversation partners from industry and the academic world from which we received many suggestions.

Münster and Dresden, April 2001

Jörg Becker
Wolfgang Uhr
Oliver Vering

Contents

1 Market and Competitive Conditions in Trading with Consumer Goods

When you look for the significant trends in trading with consumer goods, the trend to size may well initially predominate. On the one hand, mergers and acquisitions of other companies mean that the retailing companies themselves become ever larger – the growth of the German Metro, for example, is primarily based on acquisitions, but also the mergers and mutual purchases of large department stores show this – on the other hand, the sales areas are also growing. Large self-service department stores are being built on greenfield sites, specialized shops, such as do-it-yourself centers (e.g., Homedepot), are being established successfully, and also department stores strive for size. Retailing chains are increasingly supplanting independent retailers. The trend to shopping centers, which has long been established in the USA, is also coming to Europe. One of the largest centers in the USA is the Mall of America in Bloomington, Minnesota. The Mall of America, which was opened in 1992, today includes over 520 stores, employs more than 12 thousand people and has between 35 to 42 million visits yearly – more than Disney World, Graceland and Grand Canyon have together (www.mallofamerica.com). The most prominent German example is the CentrO in Oberhausen that opened on September 12, 1996. 200 businesses offer their goods under a single roof in this shopping mall with a sales area of 70,000 square meters. A car park with space for 10,500 vehicles surrounds the shopping complex and ensures problem-free logistics.

The Internet and electronic commerce form a further trend. Increasingly, products can be ordered with just a click of the mouse from the personal computer at home. In principle, anything can be bought over the Internet: from books, CDs, travel and cars, through to jam and fresh vegetables. Many pundits who see the future of buying in the Internet, predict an elimination of the stationary retailers and foresee a direct logistics chain from the producer to the final customer.

However, counter currents are becoming apparent. The corner shop that had encountered a drastic fall in sales and was considered to be dying out has experienced a renaissance in the form of service station shops (now with professional management and completely organized logistics). The ease of shopping just around the corner has also experienced a comeback, as has saving at any price at food discount stores or in self-service department stores. A relativization of the absolute pursuit of size shows itself in the shop-in-shop concept, which, within a larger unit, promotes the "familiarity" of a small, sometimes independently operated unit.

Almost certainly there will be a number of purchases that the end consumers perform with electronic commerce and the Internet. However, this new medium will establish itself in *addition* to existing shopping possibilities; a comprehensive substitution effect will occur only in a few cases. Amazon.com, the classic example of electronic retailing, started as book seller with gigantic growth rates, however, in the meantime, also sells music albums, video films, pharmaceutical products, pet requirements, second-hand books and music, greetings cards, food, toys, electronic equipment and sport articles, to mention only those areas that have been added between June 1998 and July 1999. However, the growth rates for book sales have fallen. The proportion of book sales over the Internet in the United States, the pioneer in electronic commerce, is just 3 percent. 97 percent of books are still sold in book stores or over the traditional mail order. The figures for Germany (1999): traditional book sales 89 percent, classic book mail order 7 percent, book clubs 3.5 percent and Internet book sales just half a percent. In the meantime it is assumed that a large part of electronic commerce will not take place in the business-to-consumer area but in the business-to-business segment, namely in business between companies.

These observations show: There is no clear trend in the consumer goods trading. A focussing of specific types of selling or business segments is not taking place, quite the reverse: The range of retailing forms is increasing and none of the forms appears to have long-term advantages. This results in one paramount requirement for retail information systems: the need for flexibility. Consequently, a (standard) retail information system must be capable of handling many characteristics of retailing companies as shown in Figure 1-1.

The large range of functions and flexibility of standard application systems in retailing is necessary because some retailing companies (at any one time) exhibit several forms of a characteristic and over the course of time change their characteristic forms. Thus, it is not unusual that a single company has wholesalers and retailers, stationary and mail order trading, active price policies in one range of goods and a passive price policy in another. And, in the course of time, in particular in this company, electronic commerce over the Internet and goods distribution of music albums and videos over this medium complements the classic structures. A widening of the range can be found, for example, at IKEA (augmenting the furniture range with a wide range of day-to-day products) and even the German ALDI, the classic provider of "restricted range of products and producers", has added refrigerated and frozen goods to its range of dry goods. An expansion of the logistics and customer contact can be observed for many grocery retailers, which, in addition to the classic branches, also offer Internet sales with direct delivery (e.g., Albertsons (www.albertsons.com), Asda (www.asda.com). Also the European number one in the drugstore sector, the German Schlecker, with the Schlecker Home Shopping offers Internet sales with direct delivery service for Germany, Austria and The Netherlands (www.schlecker.de).

Characteristic	**Characteristic form**				
Business level	Retailer	Wholesaler			
Extent of trading	Within the country	Outside the country			
Horizontal cooperation	Retailers	Wholesalers	Other cooperation		
Vertical cooperation	Retail and wholesale	Wholesale and industrial companies	Retail and industrial companies	Retail, wholesale and industrial companies	
Contact orientation	Stationary	Itinerant	Mail order		
Sales contact form	Sales person	Self-service	Catalog	Vending machine	
Beneficiary	Investment goods trade	Consumer goods trade			
Range extent	Wide and deep range	Wide and shallow range	Narrow and deep range	Narrow and shallow range	
Price policy	Active	Passive			
Purchase initiation through	Visit to store	Letter / fax	Telephone (caller center)	Internet	Push (e.g., Clubs)
Logistics handling	By the customer (collect)	By the retailer / intermediary (delivery)	Through the Internet (for digital products)		

Figure 1-1: Characteristic forms of retailers

The realization of an adequate data processing support for all characteristics listed in Figure 1-1 represents a major challenge for providers of retail information systems. In addition to the various forms of retail companies, in recent years various wide-ranging developments, such as the increasing decentralization of the merchandise management, the expansion of merchandise information systems to become decision support systems or the trend to point-of-sales acquisition (cf. Ahlert 1997, pp. 31), must be taken into consideration. Furthermore, a retail information system must have a structural form that allows it to be adapted to future requirements with an acceptable effort.

SAP offers with the SAP Retail system a solution that claims to be able to represent a large part of the required characteristics. We show in the following chapters what requirements the modern retail management places on information systems and how SAP has realized these requirements in their SAP Retail system. First, we derive an assignment frame that structures the tasks of retailers and which also serves as navigation aid through the retail information systems.

2 Architecture of Integrated Information Systems in Retailing Companies

Architectures, also known as grouping frames, have achieved great importance in providing an understanding of information systems at the process design level. An architecture structures the considered section of the real world, here the tasks of retailing companies, at an abstract level using the selected structuring paradigm and so emphasizes relationships and possible time-related sequences between the independent parts of the frame. The architecture of integrated information systems in retailing companies forms a pictogram of an H that represents the principal tasks. The SAP Retail system generally follows the retailing H-architecture, although not in the graphical preparation of the H. To derive the H, let us consider the principal tasks involved in retailing companies:

The goods-related requirements planning, logistical and billing functions ensure that goods flow from the supplier to the retailing company, and from the retailing company to the customer (Section 2.1). The business-administrative tasks consider the value-related sphere of the company in the cost accounting from the internal viewpoint, the billing functions of the external accounting and all tasks associated with personnel accounting, employee information, career planning and employee administration (Section 2.2). In addition to these tasks of the operative business, tasks arise form the tactical and strategic business, controlling and company planning (Section 2.3).

Interfaces to upstream and downstream operations in the same company, and to suppliers and customers, must ensure the complete, integrated, consistent and fast communication with the marketing partners (Section 2.4). The form of these tasks depends on the types of retailing business. The classic warehousing business covers more tasks and so also requires more functionality from the merchandise management system compared with the third-party business or the pooled payment business. Special sales promotions and service business provided in retailing require other functions (Section 2.5).

2.1 The Merchandise Management System

Greatly simplified, a retailing company must perform three principal tasks: procurement, storage and distribution of goods. The trading objects are goods, namely physical products. The terms service companies, brokers or agents are used when non-physical products are involved. The actual manufacturing of products does not belong to the task scope of retailing; in this case, the term industry company would be used. However, there are companies that call themselves retailing companies (retailing in the institutional sense) that perform functions beyond retailing in the functional sense (procurement, storage and distribution). Similarly, there are retailing companies that do not perform certain retailing tasks in the functional sense. An example of the former is a retailer entering service business, examples of the latter are the third-party business and the pooled payment business.

Merchandising thus covers the goods-oriented planning tasks, the logistical and the billing-related tasks that a retailing company needs to perform.

The *Merchandise Management System* is the information system that supports and controls the goods-oriented planning tasks, the logistical and the billing-related tasks of a retailing company using value and quantity-related goods transaction data.

When we consider the central tasks of retailing in more detail, we can identify the merchandise management subtasks below:

Contracting

Contracting makes the basic procurement decisions and updates the relevant base data. Central tasks are determining the suppliers with which the retailing company will enter a business relationship, determining the goods to be obtained from these suppliers, negotiating the price and conditions framework for these goods (often in annual meetings) and possibly determining the value and quantity contracts or delivery schedules.

Purchasing

Purchasing involves the placing of orders by determining the quadruple: supplier, article, quantity, time. It also includes as subtasks, the limit calculation, the requirements calculation, the purchase order quantity calculation, the stock allocation, the order transfer and the order monitoring.

Goods receipt

Goods receipt is the quantity-related logistical equivalent to the purchasing order. The goods receipt must be planned; this requires notifying the supplier so that a ramp assignment planning can be performed. Goods acceptance, sometimes also without having been ordered, and goods control, both quantitative and qualitative, i.e., related to the state of the goods, follow. The goods receipt also covers the physical goods storage, the recording the goods receipt and the analysis of the delivery notes with the assessment of the goods receipt and the updating of the goods inventory, and possibly also the handling of returns to the supplier. The administration of reusable packaging and goods on deposit are also tasks of the goods receipt.

Invoice verification

The value equivalent to the goods receipt are the invoice arrival and the invoice verification with the subtasks: invoice acquisition, invoice checking, invoice release, subsequent invoice processing and processing of subsequent conditions. On the one hand, the invoiced quantities must be compared with the order (the request to deliver), the delivery note (the delivered quantity from the supplier's viewpoint) and the goods receipt slip (the delivery quantity from the retailer's viewpoint), and, on the other hand, the invoiced values must be compared with the price and conditions scheme.

Creditor accounting

The major task of the creditor accounting is handling payments, i.e., the payment for the open items resulting from the supplier's invoice. This requires updating the creditor master data. Unless the invoice is passed automatically from the invoice verification, an entry is made for the invoice (e.g., material cost entry). Credit notes and subsequent billings may need to be booked. The payment can take place with an automatic payment run or through manual payment. If the creditor is also a debtor (e.g., for subsequent payments), the creditor accounting may also include reminder notices and interest calculation tasks.

Marketing

In the context of monitoring the retailing processes and tasks, only operative marketing is considered here as marketing; strategic marketing is considered as being a task within the company planning (see Meffert (1998) or Lilien/Kotler/Moorthy (1992)) for a comprehensive description of the marketing tasks). Here, subtasks of the (operative) marketing are the updating of customer master data, the assortment and product policies (including goods planning), in particular, assortment planning, sales planning and turn-over planning, as well as article lists. These subtasks require a customer grouping and a time-dependent article-customer assignment. The sales condition policy and the sales advertising also belong to the marketing.

Sales, goods issue, billing and debtor accounting are analog to the corresponding tasks at the goods receipt side (contracting, purchasing, goods receipt, invoice verification and creditor accounting).

Sales

Sales includes the subtasks of customer query processing, customer offer processing, creation of order records, order processing, possibly customer complaints processing, and, finally, sales representative support with customer contact, sales support and sales processing by the sales representatives.

Goods issue

Subtasks of the goods issue involve the route planning, planning of the order picking, the actual order picking, the goods issue acquisition – either at the (central) warehouse or in the branch – and entering the inventory. The shipping activities and possibly the processing of customer returns and the management of reusable packaging also belong to the goods issue.

Billing

Billing tasks include, in particular, the evaluation of the customer delivery note, the various forms of invoicing the customer (e.g., individual invoices or collective invoices) and the calculation of subsequent reimbursements, together with the production of any required credit notes and debit notes.

Debtor accounting

The central task of the debtor accounting is the administration of the debtor accounts and the monitoring of the payment. The major activities are updating the accounts receivable master data, the booking of invoices / credit notes / subsequent sales conditions, and the collection including the debiting, the booking of the payment arrival and possibly the handling of reminder notices, and the credit management (determination of the credit limit and monitoring of the credit worthiness).

Warehousing

Warehousing performs the bridging function between the procurement side and the sales side. It does this with regard to time, quantity and logistics. Warehouse subtasks involve the updating of the warehouse master data, stock transfers and posting transfers, the stocktaking in the warehouse or in the branch, and the warehouse control.

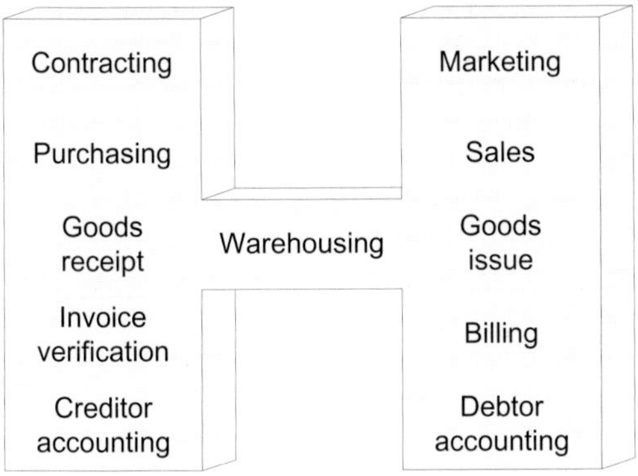

Figure 2-1: Retailing H structure

When all tasks concerned with the suppliers are placed on one leg, all tasks associated with the customers are placed on the other leg, and the logistical functions with goods receipt, warehousing and goods issue are arranged horizontally, the merchandising management-related areas of a retailing company then form an H (see Figure 2-1).

The mentioned tasks of procurement, distribution and logistics occur not only in retailing companies, but also in industrial enterprises. However, retailing has some unique features that affect the form of the tasks and consequently the information systems.

- *Contracting:* Comprehensive requirements arise for the central object of any merchandise management system, namely the article. The way of establishing the article, which, because of the large number of articles, should be as simple as possible, the time dependency of the article master data resulting from special promotions of limited duration or prices and conditions, the multi-stage organizational structures typical in retailing (in particular with regard to purchasing bureaus), and the determination of the sources of articles are unique to merchandise management systems.

- *Purchasing:* The multi-stage form of retailing companies largely determines the operative procedures. For example, purchasing must take account of the decentral and central order quantity planning. Thus, pull strategies, in which the sales at the cash desk form the starting point for requirements planning at both the branch and the central warehouse, represent a specific requirement of multi-stage retailing.

- *Goods receipt:* Whereas inventories at the wholesaler or in the central warehouse of a multi-stage retailing company are listed based on quantity, this is often not possible at retailers because of technical and organizational restrictions. To control the assortment, the article stocks are listed value-based with purchasing and selling prices at the merchandise category level (or some similar grouping). The difference between the purchasing value and selling value corresponds to the gross profit of the retailing company and provides the basis for determining the retailing margin (see Section 5.1).

- *Invoice verification:* The number of invoices to be processed means a high division of labor in the invoice verification. Accordingly, the invoice verification is normally separated from the financial accounting in retailing. The delivery notes and the invoices (normally only the invoice total) are input in separate work steps. The invoices are then automatically matched against the processed delivery notes.

- *Warehousing:* The warehousing functions in retailing are especially dependent on the retailing level. Retailers normally value the stock based on the selling price. The handling of seasonal articles in the central warehouse for assortment wholesalers at the end of the season requires the transfer between the picking and the reserve warehouse areas. There is also a change of the assignment for fixed picking locations to articles, because, for example, summer season goods are no longer stocked in the picking warehouse in the fall.

- *Marketing:* The articles must be assigned to branches or customers in the marketing area (so-called listing). The article assignment is the result of policies based on the assortment. The large number of articles and the heterogenity of the selling means that classifications of articles and customers must be made in order to achieve the fastest and most efficient assignment of the articles to customers. A further characteristic within the marketing tasks in retailing companies involves the support for special promotions. This requires not only different logistical procedures but also special billing techniques, which, for example, affect the granting of conditions in the debtor and creditor areas. The promotion business provides a suitable starting point to create processes that transcend business levels, because up to now the promotion activities from industry and the promotion activities from retailing companies lacked in the coupling of information. This means unused integration potential remains.

- *Goods issue:* Goods issue covers the logistical processes from the picking through to the shipping. The picking orders from the various picking areas must be grouped and loaded in shipping zones. The picking order normally results from the route plan for the day. The enormous fluctuations in the picking load requires that the capacity management provides profiles for capacity demand and capacity availability as used in industry. The heuristics known from industry can be used for this subproblem.

- *Billing:* Billing involves invoicing the customer for the provided services. Many wholesale companies demand support for the pooled payment business form. This requires the combining of the supplier and customer invoices.

2.2 Business Administrative Systems

The goods-oriented planning, logistical and billing tasks of merchandise processing are augmented with the business-administrative systems of the general ledger and assets accounting, cost accounting and personnel management (human resources management). Whereas the merchandise management tasks relate either to the supplier or to the customer or are of a logistical nature, business-administrative systems do not have this focus.

General ledger accounting

The general ledger accounting combines the subledger accounting of the creditor accounting, the debtor accounting and the materials accounting, and summarizes them to produce the general ledger accounts. Tasks of the general ledger accounting involve maintaining the general ledger account master data, booking of the general ledger accounts, bank transactions and preparations for financial statements, the creation of the financial statement, and the financial planning.

Fixed-assets accounting

In addition to maintaining the fixed-assets master data, the fixed-assets accounting, as another subledger accounting, includes entry bookings for fixed-assets, bookings for depreciation, entry postings, transfer postings and issue postings together with final statements and fixed-assets controlling.

Cost accounting and service costing

Whereas the financial accounting is associated with the external accounting, the cost accounting and service costing are assigned to the internal accounting. Because key figures used in the controlling build on the results (e.g., cost and profit types) from the cost accounting, it provides the information basis for the controlling. The cost accounting tasks involve the master data administration with the management of supplementary cost and profit types, the management of the reference quantities and the reference objects, in addition to the cost and profit planning with the planning of the sales proceeds, the goods resource costs and the costs of inner-company services, the acquisition of the actual data and the analysis and monitoring, in particular profit margin calculations and business comparisons. Central profit margin calculations in retailing are the merchandise category profit margin calculation, the supplier profit margin calculation and the customer profit margin calculation.

Human resources management

In addition to various accounting tasks, the human resources management also includes planning tasks. Maintaining the personnel master data covers the creation and updating of employee master data (e.g., personal data, working form, tariff group, bank account details) and the assignment of employees to organizational units (e.g., employee is purchaser for a specific purchasing organization, planning officer in a branch or a warehouse). The accounting task of the human resources management is the personnel payment. Planning tasks involve determining the personnel requirements using a personnel requirement planning, the personnel acquisition planning and possibly a personnel allocation planning, together with a personnel assignment planning and a personnel cost planning. The personnel evaluation and the personnel development planning are further associated tasks.

Because the architecture of the retailing H-model groups all supplier-related tasks in one leg of the H and all customer-related tasks in the other leg, the creditor and the debtor accounting are on the supplier-related side and on the customer-related side, respectively. This also agrees with the definition of a merchandise management system that combines the goods-related planning, logistical and billing functions. In contrast, the grouping of the bookings in the general ledger accounting is assigned to the foot of the H. However, in many data processing solutions, the creditor accounting and debtor accounting are integrated with general ledger and assets accounting in the financial accounting system. This is also the case in the SAP system, where the goods-related planning and logistical tasks belong to the SAP Retail system, and all billing tasks belong to the SAP R/3 FI financial accounting module.

2.3 Controlling and Company Planning

Whereas the previously listed tasks have a more operative nature and describe the day-to-day retailing business, controlling and company planning involve the strategic tasks of retailing management. For support purposes, the data are required in a highly aggregated form that supplies the important operating numbers. These then form the basis for company decisions. The aggregated data are extended with external data, e.g., from market research companies such as GfK or Nielsen. Technically, data warehouse systems, large databases that are optimized for rapid data retrieval and forming operating numbers for large data volumes (also refer to Chapter 5), are often used for this area.

The difference between controlling systems, executive information systems and systems for company planning is somewhat fuzzy. The tasks that arise here are cyclical analyses for price and sales controlling or logistics controlling and profit-related analyses, such as controlling specific sales promotions or determining the advertising success, that can be assigned to the controlling, operations type and competition comparisons and market data analyses that can be more appropriately assigned to the executive information systems area, and tasks of the business area planning, location planning and strategic logistical planning that belong to the company planning.

2.4 The Retail Information System and the Retailing H-Modell

As shown before, a retailing company has to perform not only merchandise management tasks, but also business administrative tasks as well as general controlling and company planning tasks. Because of this a merchandise management system does not offer all required IT-support. Additional information systems and applications are necessary. An information system, which covers all these tasks, we call *retail information system*. Logically and maybe also physically the merchandise management system is part of the retail information system. SAP Retail does cover more functionality than only the merchandise management, so it is a retail information system.

In the following we generally use the term retail information system, just to emphasize that only merchandise management tasks are meant, we fall back to the term merchandise management system.

The variety of the retailing tasks and so the functionality of retail information systems can be structured based on the retailing H structure. But there are different viewpoints, which have to be considered.

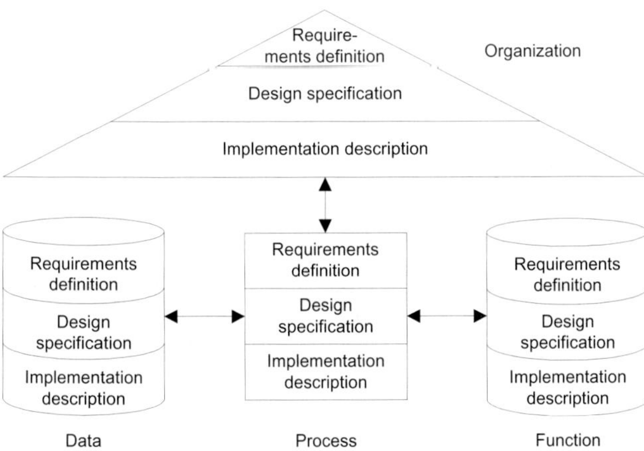

Source: Scheer (2000), p. 1.

Figure 2-2: ARIS architecture

In accordance with the ARIS architecture (architecture of integrated information systems, see figure 2-2) from Scheer, a differentiation can be made between *functions* within these tasks, namely the subtasks to be performed, *data*, namely the structure of the tasks, and *processes*, namely the behavior of these areas (cf. Scheer 1999, Scheer 2000). Where, however, function and process are not completely distinct. The process contains, in particular, the temporal sequence of individual transactions, alternatives in the procedure and parameters on which these alternatives depend.

The retailing H-model summarizes the tasks of a retailing company and the different viewpoints (see Figure 2-3). At first glance, it would appear that the retailing H-model applies primarily to wholesalers in which the tasks on the supplier side and the tasks on the customer side have similar structures. In contrast, stationary retailing involves some tasks from the right hand side at the point of sale, i.e., at the cash desk. This produces a reduced retailing H model.

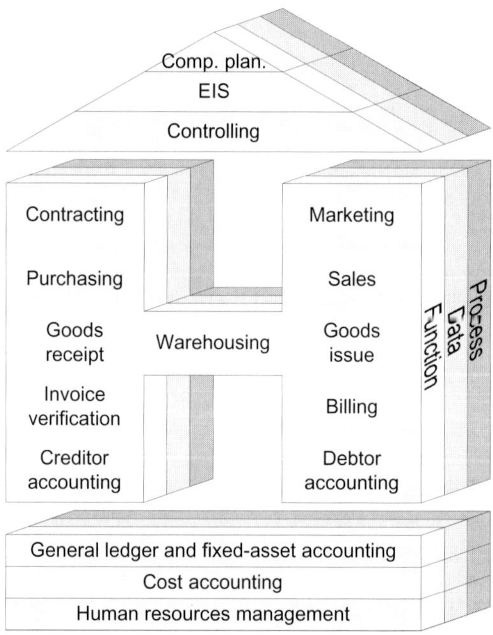

Figure 2-3: Retailing H-model

However, the retailing H model for retailers shown in Figure 2-4 applies only to "classic" stationary retailing. If the retailing company issues customer credit cards, then the retailer knows the debtor, and billing and debtor accounting become evident again as tasks. Mail order (also a form of retailing) and Internet trading require all functions of the complete retailing H model. Thus, the retailing H model covers a superset of the tasks of retailers and wholesalers. It can be "shortened" appropriately for specific situations.

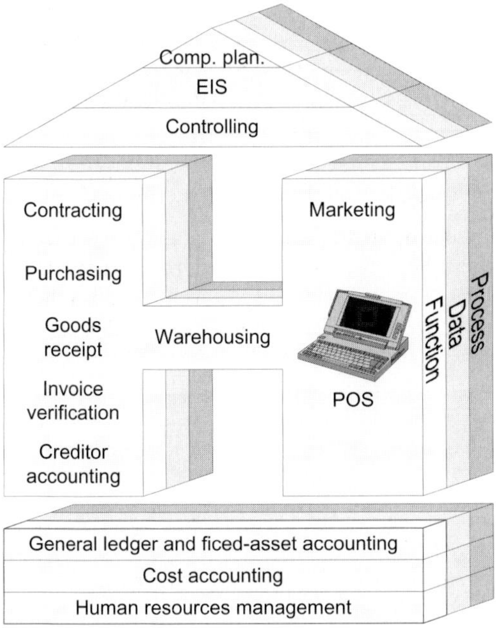

Figure 2-4: Reduced retailing H-model for retailers

2.5 Intra-Company and Inter-Company Integration

In addition to the integrated and efficient handling of tasks within a company, the integration and efficiency of handling the intra-company and inter-company processes are increasing in importance with regard to Supply Chain Management and "Efficient Consumer Response". These include, on the one hand, the planning, logistical and billing tasks from head office to the branches or from the cooperative head office to the independent retailers or from franchiser to franchisee, and, on the other hand, the relationships to "external" suppliers and customers. The development starts with the bilateral automation of the business correspondence, includes standardization of the communications data as part of SEDAS and EDIFACT, and finishes with the Internet-based electronic business correspondence and automated payment (also refer to Chapter 7).

2.6 Types of Business

The classic type of retailing business is warehousing business with the procurement, storing and distribution functions described in the retailing H model. In addition to the classic warehousing business, third-party business and the pooled payment business have established themselves, and orthogonal to these, the promotional and service types of business that can occur in conjunction with the first three types of business.

Third-party business

For classic third-party business, a customer places orders with the retailing company, which forwards the order to the supplier; the delivery is made directly from the supplier to the customer. The supplier invoices the retailing company, which pays; the retailing company independently invoices the customer (see Figure 2-5). The logistics take place directly between supplier and customer. The associated retailing H model for classic third-party business reduces to the functions shown in Figure 2-6.

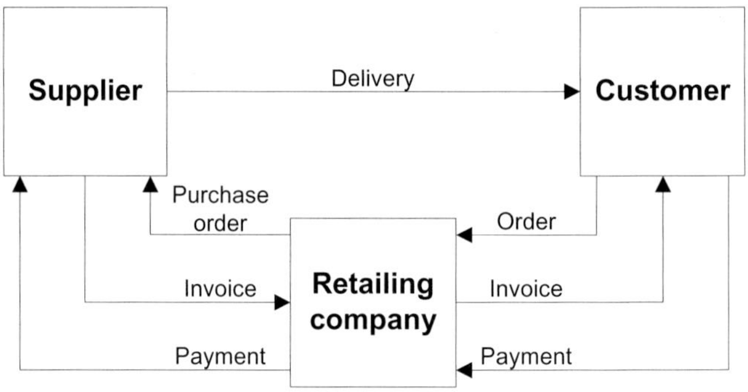

Figure 2-5: Classic third-party business

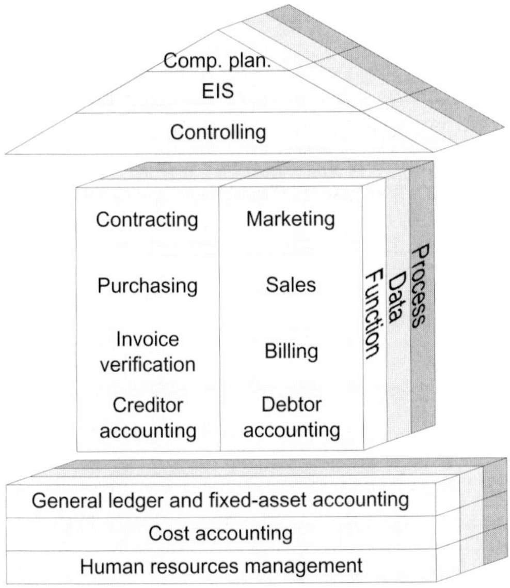

Figure 2-6: Retailing H-model for third-party business

A second form of the third-party business occurs in retailers with branches. The branch orders directly from the supplier (sometimes without entering an order in the information system). The supplier then delivers to the branch and sends the invoice to the head office (and to the branch) (see Figure 2-7). The head office is responsible for monitoring the process. This transfer handling proceeds similarly to the first case when the orders are entered in the system and the transfer is made centrally to the supplier.

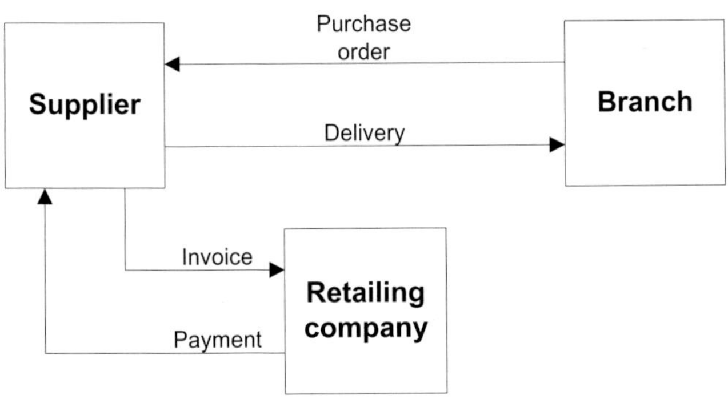

Figure 2-7: Store-based third-party business

Pooled Payment business

The tasks that the retailing company needs to perform for pooled payment business reduce still further (see Figure 2-8). The handling of both the logistical and the planning functions takes place directly between supplier and customer, whereas the retailing company is involved only with the billing tasks.

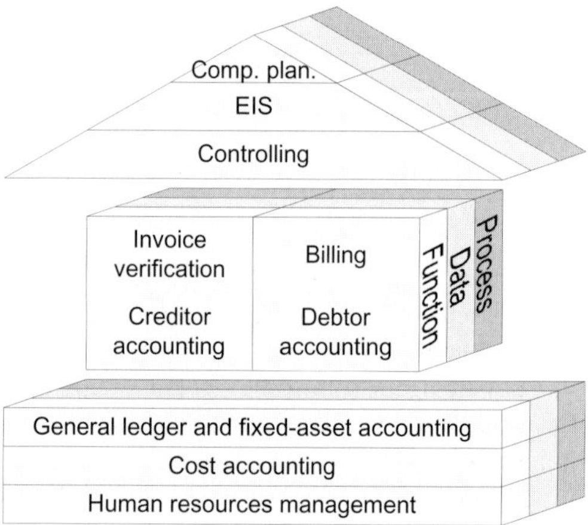

Figure 2-8: Retailing H-model for pooled payment business

The customer places orders with the supplier, which then delivers directly to the customer who also receives the invoice from the supplier. The supplier sends a copy of the invoice to the retailing company. The retailing company pays the invoice amount and sends an account statement for payment to the customer (see Figure 2-9). Because the creditor invoice flows immediately into a debtor account statement, the creditor and debtor accounting are closely connected. The master data for creditors and debtors must also be closely coupled, when, for example, the creditor has bilaterally agreed specific terms with the debtor (payment terms, immediate rebate).

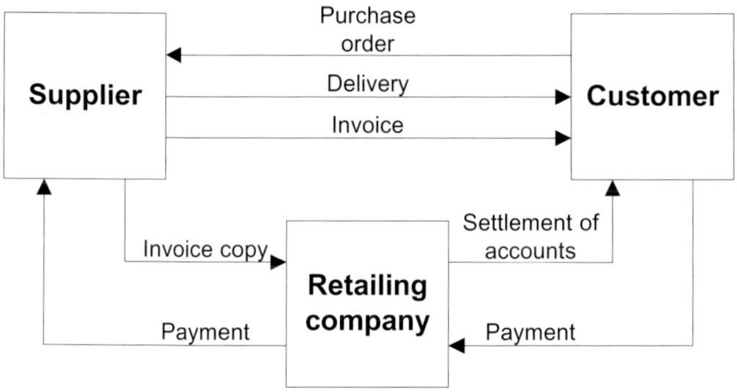

Figure 2-9: Pooled payment business

In the case of pooled payment business, the retailing company normally provides a default guaranty for its customers' purchases (acceptance of the factoring discount). The retailing company for the pooled payment business acts primarily as bank. It acts as the paying agent for the customer's debts. The specific central settlement contract can be either a contract with guaranty character or a form of *factoring*. Factoring here is understood to be the contracted continuous purchase of debt claims (normally before they become due) from deliveries and services through a factor while assuming certain service functions and often also the risk of default (cf. Perridon/Steiner 1995, p. 401). The feature for the handling of the pooled payment business is the coincidence of invoice payment and billing (in the sense of presentation of the invoice). In the legal sense, the pooled payment business transaction is *one* business transaction, whereas the third-party business represents *two* separate business transactions.

Promotion business

Promotion processes and service processes can occur during warehouse business, third-party business and pooled payment business (for detailed information about promotions refer to Merrilees, Miller 1996, p. 261f.). Promotions are purchasing and/or sales measures that take place over a limited time and make use of conditions and other terms of the suppliers and/or serve to increase sales. Especially in the consumer goods retailing, and in particular in the food retailing, sales promotions have a very large importance. To attract off-the-street customers, the prices of a range of products are reduced, sometimes drastically, in periodic intervals, sometimes even weekly. In the promotion business, the tasks of the sales side are more closely coupled with the tasks of the purchasing side than is the case of traditional warehousing business. If a promotion is planned, the head office negotiates with the branches the quantities of the promotion articles to be ordered. This allows a preorder to be defined. This permits the head office to define the quantities of the promotion articles to be purchased. These quantities then serve as the basis for the price negotiations with the suppliers. In this case, there is no normal stocking in the warehouse, but often a direct transfer from the goods receipt to the goods issue (transshipment or cross-docking, see section 3.3). Because the promotional goods are billed separately, this requires a time-related control of the prices, a non-trivial problem for the merchandise processing systems.

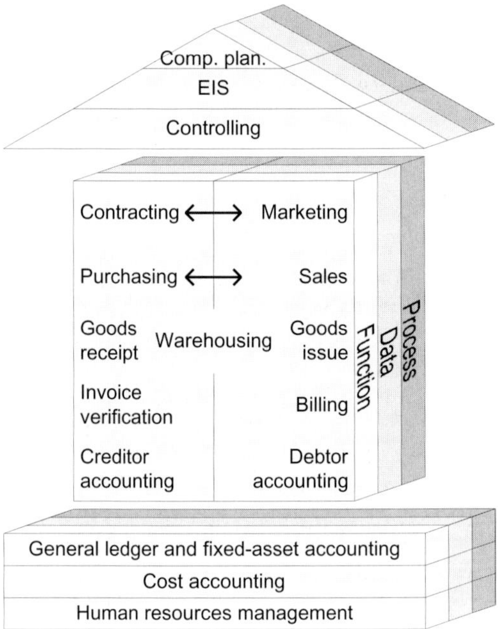

Figure 2-10: Retailing H-model for promotion business

Figure 2-10 shows the typical retailing H-model for the promotion business in which the coupling of the procurement and sales sides causes the two sides of the H to move together.

Service business

Service business completes the activities of retailing companies. The head office / wholesaler / franchiser perform here services for the branch / retailer / franchisee with regard to business and tax consultancy, data processing consultancy, marketing support, recommendations for entering the market or provision of system solutions.

3 Closed Merchandise Management Systems and Their Realization with SAP Retail

Chapter 2 described which functional subareas a merchandise management system must always cover. This chapter considers the individual subfunctions in more detail and presents the concepts realized in SAP Retail.

The following sections assume a *closed merchandise management system*. A *merchandise management* system is called closed (with regard to the goods cycle), when the goods inventories in all company units (in particular, also in the decentral units, namely stores) are maintained for each particular article and the article stocks are updated in real-time (cf. Hertel 1999, p. 6 f.; Ahlert 1997, p. 34 f.). The latter implies that the goods arrivals, the goods losses caused by spoilage and breakage, and sales, for example using scanners at the cash desks, are recorded for each particular article. In "non-closed" *merchandise management* systems, the goods cycle is "closed" only after the annual (manual) physical inventory has been performed.

However, the advantage of closed *merchandise management* systems, which for quite some time represent the state-of-the-art, is not only of a qualitative nature with regard to the question when the system can be "closed" and so the article inventory is updated. The expanded possibilities that arise from the current inventory information are more important. In particular, the ability to use automatic purchasing or article-specific (short-term) profit margin calculations.

Before we discuss the operative functions, the following section first introduces the central organizational structures and master data objects. The consideration of the operative functions then follows the structure of the retailing H model presented in Chapter 2.

3.1 Organization Structures

The representation of organization structures in information systems forms a fundamental prerequisite for their use in mid-sized and large retailing companies. The multi-stage structures typical in retailing cover the industrial production, include wholesalers, possibly intermediate stores and individual retailers, and end with the consumer. Such structures have a complexity that places high demands on an integrated information system. Figure 3-1 shows the basic structure of a multi-stage retailing company that provides both wholesaling and retailing functions, and which is present in the market with several sales lines.

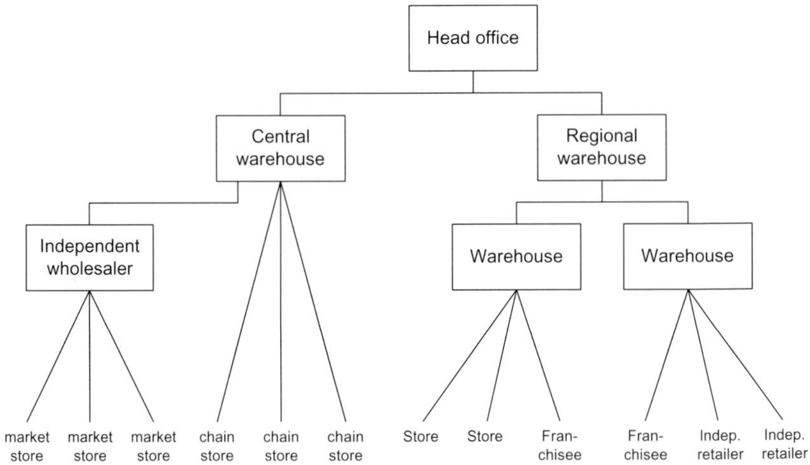

Figure 3-1: Typical structure of a multi-stage retailing company

However, such a one-dimensional representation cannot adequately describe the organization structure of a complex structured retailing company. Rather, the different views of the organization structures must be differentiated. The *legal viewpoint* can be described with details of the legal forms of companies and their owner structures. In contrast, warehouses, storage locations, unloading and receiving points are relevant from the *logistical viewpoint*. A (largely) arbitrary relationship can exist between a warehouse (logistical view) and a company (legal view). In a simple case, a warehouse belongs to just one legal company and could possibly be at the same site; in more complex situations, parts of a warehouse can belong to several companies and can be located geographically at a different location.

There are also a *buy-oriented view* that describes the organizational responsibilities in the purchasing department and an analog *sell-oriented view*. Typical buy and sell-oriented organizational units are purchasing and selling organizations, distribution channels or purchasing areas. The following sections discuss these in detail. A further central view is the *goods-oriented view*. This consideration focuses on the article and article groupings (merchandise categories and assortments).

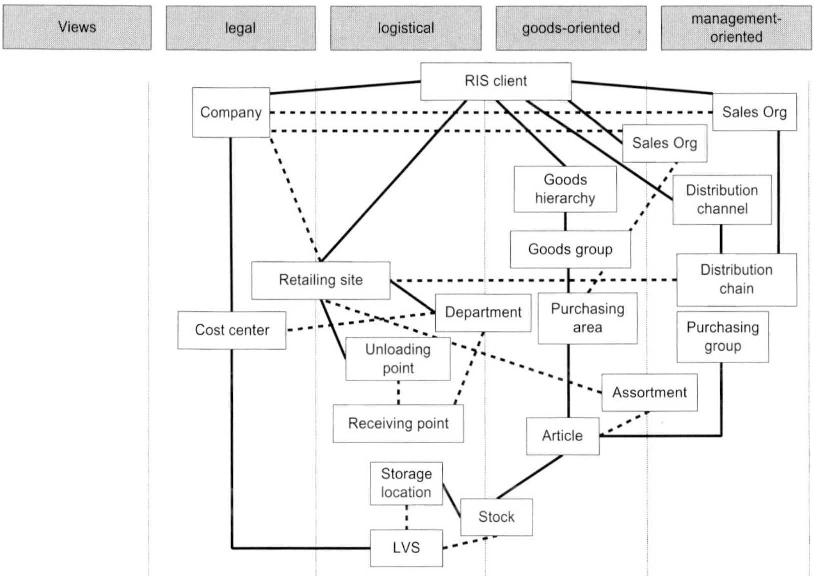

Figure 3-2: Organization structures in SAP Retail (selection)

Figure 3-2 shows the most important organization structures (grouped according to the appropriate views) in SAP Retail. The buying-side and selling-side views are grouped in this figure as a management-oriented view.

Although such a semiformal representation can illustrate basic interrelationships, the semantics of the displayed relationships are often unclear. A unique, formalized representation of the organization structures and the relationships that exist or could exist between the individual organization structures can be obtained from a *data model.*

A usual form for data models at the process design level is the entity-relationship model (ERM) developed by Chen (Chen 1976, pp. 9-36). SAP has developed SAP-SERM, its own data modeling method based on ERM. An explicit specification of various relationship types and a simplified representation of the relationships compared with the ERM often produce more easily understandable and more compact models compared with the ERM representation. The following discussion introduces the major constructions and basic rules of the SAP-SERM notation (consult Becker/Schütte 1996, pp. 37-47 for a more comprehensive explanation).

Discussion: Data Modeling with SAP-SERM

SAP-SERM uses entity types, semantic relationship and specialization types as central modeling objects. SAP-SERM does not explicitly display in graphic form the relationship types known from the ER model.

Entities represent real world objects that are important for the company. However, the individual objects are not modeled, but rather only object types (*entity types*). For example, the customers "1235 Brown Company" and "1563 Smith and Sons" (and all others) are grouped as the Customer entity type.

The *relationship types*, which describe the relationship between two entity types, have particular importance in the SAP-SERM method. Relationship types are described with a cardinality that specifies the numeric relationship with which the entities can be associated with each other. A 1:1 relationship between an entity type A and an entity type B means that just one entity from B exists for each entity from A. In contrast, a 1:m relationship allows several B entities to exist for one A entity. Different graphical representations are used in SAP-SERM to illustrate the cardinality (refer to Figure 3-3).

Graphical representation	Designation in SAP-SERM	Designation in ERM
——>	1 : 1	(1,1) : (1,1)
——+>	1 : C	(0,1) : (1,1)
——>>	1 : M	(1,m) : (1,1)
——+>>	1 : CM	(0,m) : (1,1)
C = conditional; m/M = many		

Figure 3-3: Notation rules for relationship types in SAP-SERM

The entity types are arranged in SAP-SERM so that a left-to-right edge (relationship type) represents an existence dependency. Thus, in the model, entity types positioned further to the right depend on one or more entity types arranged further to the left.

SAP-SERM differentiates between three basic types of relationship: hierarchical, aggregation and referential relationship type. To simplify the recognition of the relationship type, the edges are labeled with an alphabetic character. H, A and R represent a hierarchical, aggregating and referential relationship type, respectively. (In addition, a differentiation is also made between conditional-aggregating (CA), conditional-referential (CR) and referential-temporary (RT) relationship types; these represent special forms of the three basic types.)

Figure 3-4: Hierarchical relationship type

For the hierarchical relationship type (H), one entity type depends on another in a way that the dependent entity type can only be defined and assigned through the existence of the other entity type. A hierarchical relationship type represents a semantic refinement. The existence dependency specifies that the dependent object (e.g., warehouse area) cannot exist when the higher-level object (e.g., warehouse) no longer exists (see Figure 3-4).

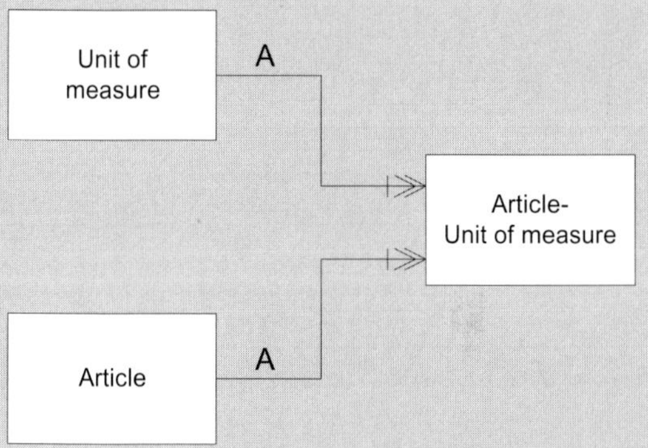

Figure 3-5: Aggregating relationship type

The *aggregating relationship type* represents a relationship between two or more entity types (see Figure 3-5). Thus, the relationship between articles and the generally defined units of measure describes the article units of measure, i.e., the units of measure possible for the concrete article.

Considered formally, a *referential relationship type* always assigns to an entity type the key attribute of the referenced entity type as non-key attribute (see Figure 3-6). For example, a retailing promotion makes a reference to a distribution channel, namely, retailing promotions can be differentiated according to distribution channels. However, this also means that retailing promotions cannot be defined without reference to a distribution channel. The *conditional-aggregating relationship type* corresponds to the aggregating relationship type, where an entity type *can* be aggregated from another (whereas it *must* be aggregated for an aggregating relationship type). The *conditional-referential relationship type* corresponds to the referential relationship type with the restriction that the referential relationship is possible but not mandatory.

The *temporary-referential relationship type* is the weakest type of a referential relationship. It is possible that this relationship can change during the lifetime of an entity or sometimes not even be present at all.

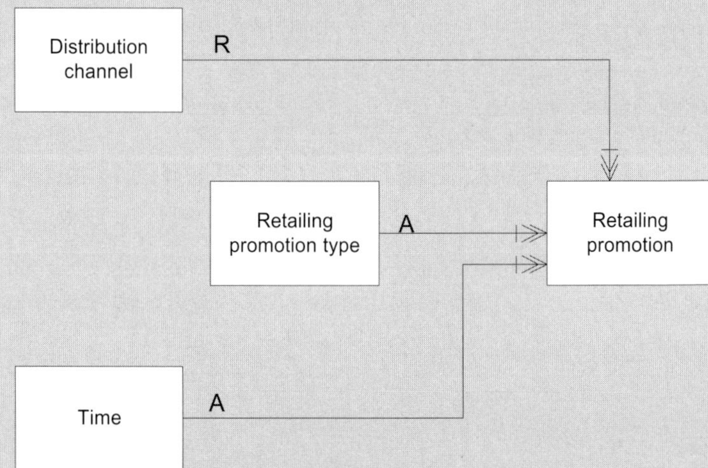

Figure 3-6: Referential relationship type

SAP-SERM uses a triangle to indicate a *specialization*, such as the differentiation of customers into end-customers and resellers (see Figure 3-7). The general entity is placed at the left of the triangle; each specialization is listed at the right-hand side of the triangle (normally below each other).

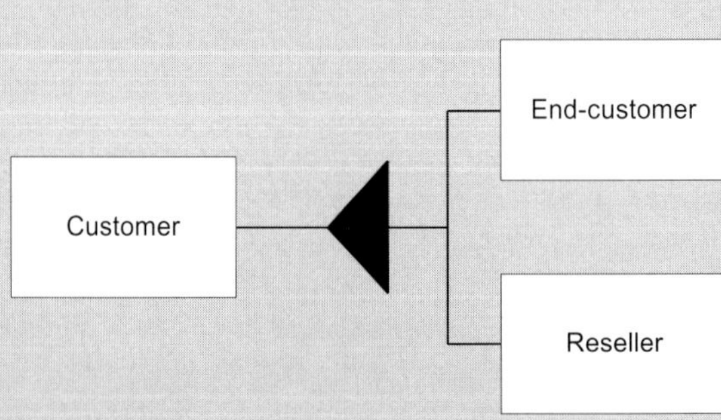

Figure 3-7: Specialization in SAP-SERM

SAP-SERM has differentiated rules for the horizontal and vertical arrangement of the entity types in the model. The central rule here is the previously mentioned horizontal placement of the entity types from left to right in accordance to their existence dependency.

Almost all large vendors of retail information systems provide data models of their system or at least for selected subareas. SAP supplies a detailed data model of the complete SAP Retail System in SAP-SERM notation. The complete model includes several thousand entity types and their relationships. Various selection and navigation capabilities allow the selection of the relevant subareas.

An explicit data modeling is an important part of the software creation to permit a data integration (i.e., support for a common creation and use of data by different areas) and ensure consistent and up-to-date data.

As part of the software selection and the software implementation, data models provide an important information source about the structural power of a retail information system. Because subsequent changes to constructions not contained in the data model affect the basic structures of the software system, such additions can be performed only with significant effort. Furthermore, such changes normally result in a loss of the release capability. This means, new versions of the software cannot be used directly, rather individual modifications must initially be made manually for each release change.

The representation of the company-specific organization structures with the structures provided in the information system represents a central step in the software implementation project that should be performed with utmost care. Important here is the appreciation of the effects of alternative representation capabilities for the company's organization structures with regard to both the later functionality and the system performance. Once the decision has been made for an organization structure, significant effort is needed to change this subsequently. Two basic concepts are possible to match the specific (historic) organization structures of the company with the structure provided in the software:

- **Adaptation of the system**

  Companies often have the wish to continue to use existing organizational structures without change when they introduce implementation projects with standard software. This should ensure the same organizational levels can continue to perform the company's business processes. Because a parallel adaptation of the company organization structure increases the complexity of the software implementation, this requirement is understandable.

  SAP Retail can be *customized* to the existing (or desired) organization of a retailing company. Before the Retail System was released, the structures were tested at many retailing companies that represented a good cross-section of the retailing industry. However, despite this high structural flexibility of SAP Retail, it is possible that small changes must be made to your own company organization structure.

- ***Adaptation of the company***

  Although an adaptation of the software is normally simpler than a change of the organizational structure of the company, it should not be forgotten that the structures in retailing companies have often grown for historic reasons and/or are the result of a previously inadequate data processing support. For this reason, the implementation of a new software should be the motivation to critically question your own structures. Any necessary organizational changes must be initiated from top management and, when required, realized directly as part of the software implementation project. This initially requires that you completely understand the structures provided by SAP Retail in their business meaning and their interactions. In a second step, a restructuring plan for the existing organization structure of the company is produced.

The knowledge of the organizational structures provided in SAP Retail and their central meaning and functionality is important in both cases.

An SAP system can be divided into several *logical systems,* which are also known as *clients*. Almost all definitions and specifications in SAP Retail are client-related, i.e., they do not apply between clients. Thus, several clients are used principally when several (not only legally) independent companies are to be represented on a single SAP system. This could be the case, for example, in an external data processing center that operates SAP systems for various companies.

The client structure can also be used to provide a company with both an operative SAP system and a test system. The data can be copied between the operative and the test system using a client copy, a comprehensive transfer of all settings and the client data.

In contrast, because a client can cover several legally independent companies (e.g., subsidiaries) that are separated from each other using so-called company codes, company structures are often represented within a client. A *company code* in SAP Retail is the smallest organizational unit of the external financial accounting for which a closed accounting that meets all legally required proofs for the individual accounts, such as balance sheets and statements of earnings, is possible.

For a group with several subsidiaries that need to use shared base data, the group then corresponds to a client, whereas the subsidiaries are represented using company codes.

The following section introduces the central organizational elements of the logistical, the purchase-oriented and the sales-oriented view with references to the SAP Retail data model. The goods-oriented aspects (article, merchandise category) are considered in Sections 3.2.4 and 3.2.5.

3.1.1 Logistical Structures

Logistical structures are used to organize a company with regard to the purchasing logistics, the sales logistics and the warehouse management. The central logistical unit in SAP Retail is the *site*. A site is an organizational unit of the logistics that divides the company with regard to the production, the inventory management, the sales or maintenance work. As retailing-specific form of the site, the *retail site* has been added to SAP Retail. A retail site can be further differentiated into distribution centers and stores. A retail site is characterized by making goods available for distribution or sales. Whereas in *distribution centers*, goods are stored for other retail sites, *stores* represent the subsidiaries in which the goods are presented and sold. Distribution centers are thus a construction to represent central or regional warehouses used to supply stores.

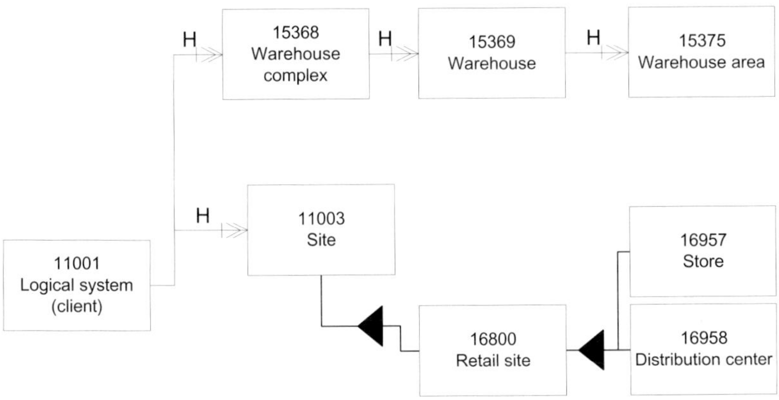

Figure 3-8: Basic logistical structures (selection)

The basic warehouse structures are represented using the organizational units: warehouse complex, warehouse and warehouse area. A *warehouse* typically corresponds to a building or a part of a building. The *warehouse complex* can be used to group several warehouses (e.g. at a company site), and organize and manage them together. The *warehouse area* divides a warehouse – primarily for purposes of stock placement or picking – into several logical subareas.

Figure 3-8 shows the structural interrelationships as SAP-SERM data model. Figure 3-32 on page 77 provides a detailed description of the internal warehouse organization (warehouse areas, storage bins, assignment of articles, etc.).

3.1.2 Purchase Structures

The *purchasing organization*, the central construct of the purchase-side organizational structures, is responsible for the negotiations with the vendors. If a purchasing organization makes cross-company agreements, it acts cross-company or even cross-group. Examples for cross-group purchasing organizations in Germany are the national purchasing offices, which, in the food retailing area, include Markant AG, the Rewe Zentrale AG and the Edeka Zentrale AG as the most important representatives. Increasingly, a connection of national offices to international offices (also known as mega-offices) can be observed. Examples for such mega-offices are the EMD, the AMS, the CEM, the EUROGROUPS and the INTER COOP. The Swiss EMD consists, for example, of the member companies Selex (Italy), Markant (Netherlands), Markant (Germany), ZEV (Austria), Selex (Spain), Sodacip (France) and Uniarme (Portugal).

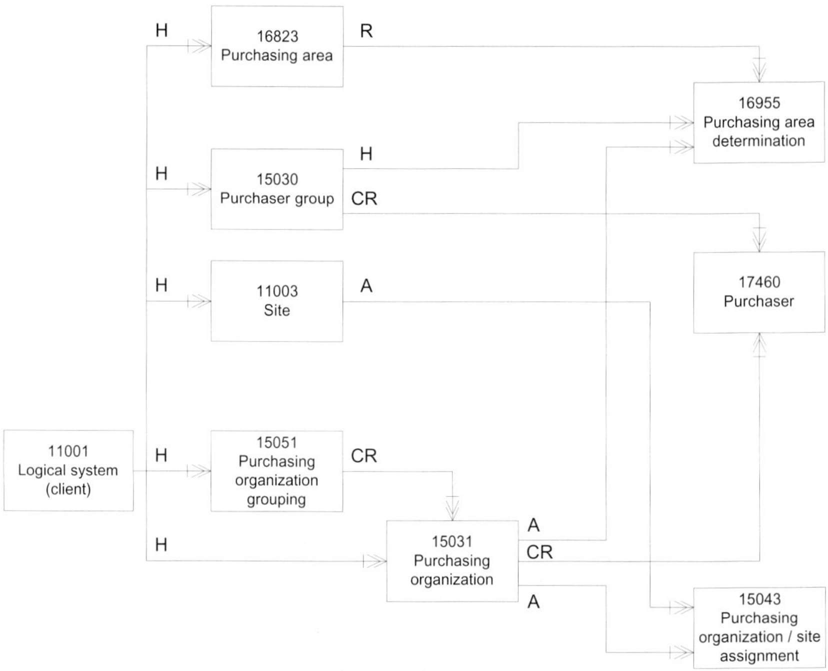

Figure 3-9: SAP data model purchasing organization (selection)

Figure 3-9 shows a selection of the most important purchase-oriented organizational structures in SAP Retail. As expected, the purchasing organization adopts a central role. For legal responsibility reasons, it is assigned to a company. The purchasing organization grouping can be used to combine several purchasing organizations to form complex (cross-group) purchasing organizational structures.

Because several purchasing organizations can be defined, it must be determined for which areas which purchasing organizations are responsible. The purchasing organization site assignment serves for this purpose.

Purchasers represent concrete employees of the retailing company to which purchase or requirements-planning relevant tasks are assigned. Purchasers can be combined to form purchaser groups that are assigned to purchasing organizations. Traditionally, purchasers in retailing are responsible for certain assortment parts and/or vendors. SAP Retail represents this with the *purchasing area*. The *purchasing area determination* specifies which purchaser group / purchaser is responsible for which purchasing areas.

3.1.3 Distribution Structures

Figure 3-10 represents the central organizational structures for distribution. Several sites can be assigned to a *company code* as a legally independent company. A single site always exists only in just one company code. One or more *sales organizations* are assigned to the company code. This is also a 1:n assignment.

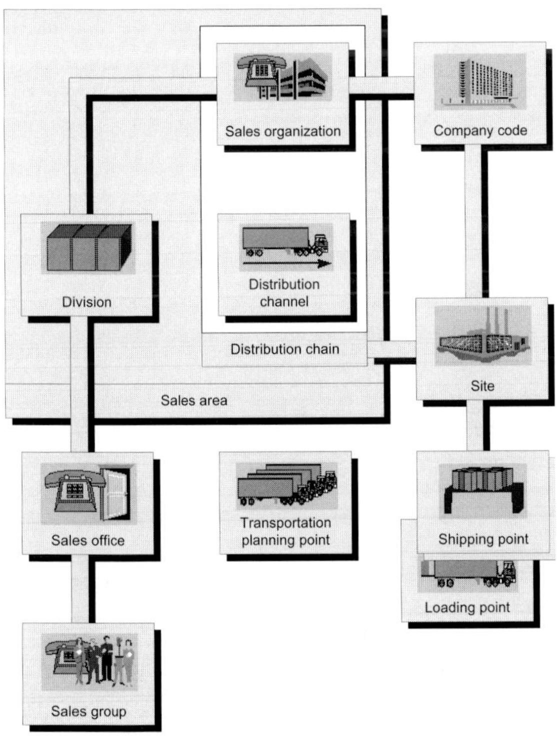

Figure 3-10: Distribution-side organizational structures in SAP Retail

The sales organization has the task in SAP Retail to further divide the sales within an organization. If this differentiation is not required, a "general sales organization" can also be used. Specific customer and article master data can be defined within a sales organization. In addition, prices and conditions can be individually defined for these articles. Together with the *distribution channel*, the sales organization forms a *distribution chain*.

Divisions, a common construction used in industry to subdivide the sales, are directly assigned to one (or more) sales organization(s) in SAP Retail. In the next step, sales areas can be represented as a combination of distribution chain and division. The system automatically monitors the consistency conditions. For example, if the sales organization 1000 ("Germany") and the distribution channel 10 ("end customers") are combined to form a distribution chain, the division 05 ("foodstuffs") can only be assigned to this distribution chain when the division 05 was previously approved for the sales organization 1000. Offices and individual employees can be assigned to sales areas defined in this way. From the sites point of view, several physical shipping points can be defined for a site. The mentioned interrelationships can also be seen in the associated data model (see Figure 3-11).

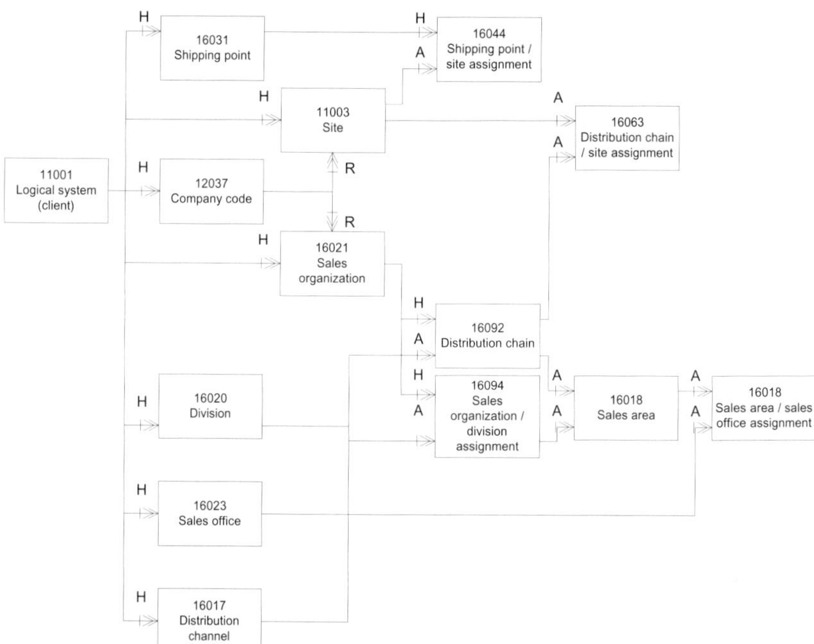

Figure 3-11: SAP data model sales organization (selection)

SAP Retail, as part of the customizing of the organizational structures, provides an automatic consistency test. This test is explained using the sales area as example: At all organizational levels in sales, the assignments are tested for logical consistency and those organizational levels are shown for which no assignments exist. A test of the errors and warnings during the implementation can determine which existing organizational relationships of the company have not yet been fully represented in the system. Any missing settings can be added later. The test should be repeated until no further errors or warnings occur.

3.2 Basic Data Management

In addition to the definition of the organizational structures, the basic data management of a *merchandise management* system forms the basis for performing all operative business processes. The basic data management covers the creation and updating of the classic master data (e.g., article data, customer data and vendor data), the merchandise categories, and the purchase and sales-side conditions.

The importance of an efficient basic data management in retailing results from two factors: on the one hand, current and complete basic data form the basis of the operative business activities. Thus, the absence of relevant article and customer master data, and stored conditions (wholesale area) means that business transactions can be handled just as inefficiently as in the retailing area without valid EAN. On the other hand, both the amount of master data and the number of business transactions in the retailing are typically significantly larger than in industry. Figure 3-12 lists sample values for technical wholesalers with their business transaction volumes (requests for quotation, purchase orders and invoices, etc.) and the number of active articles. The larger number of business transactions compared with industry (for comparable annual sales) result from the smaller average order value for each customer order, which is typical for retailing.

	Company 1	Company 2	Company 3
Company type	technical & electronical wholesaler	technical wholesaler	technical wholesaler
Annual sales	$ 85 Mio.	$ 40 Mio.	$ 379 Mio.
(Active) articles	60.000	74.000	84.000
Annual requests for quotation	120.000	2.500	k.A.
Annual purchase orders	32.000	40.000	480.000
Annual vendors invoices	25.000	58.000	420.000
Annual customer quotations	100.000	15.000	370.000
Annual sales orders	450.000	250.000	2.200.000
Annual sales invoices	355.000	100.000	1.200.000
Average sales order value	$ 191	$ 161	$ 300

Source: Schütte, Vering, Wiese (2000), p. 106.

Figure 3-12: Sample business transaction volumes for wholesalers

When we consider these data in more detail, the major importance of an efficient DP support for both the basic data management and the processing of the operative business transactions becomes apparent:

Company 3 listed in the figure handles approximately 2.2 million sales orders annually. In accordance with the usual conditions for wholesalers, these sales orders are recorded completely with customer reference. Let us assume that the order entry with good DP support can be processed only one minute faster per order, this results in an annual difference of approximately 36,700 person-hours (or more than 4,500 person-days). Similar calculations can be performed for customer quotations, requests for quotations and purchase orders.

The assortment range (for example, 50,000-200,000 active articles in the wholesale and department store area) and the dynamics of article and assortment changes (addition and discontinuation of articles, changes to the article specification, and the article prices and conditions) require a high maintenance effort.

There are various concepts to provide article master data and master data changes in digital form in the retailing industry, and so reduce the manual acquisition and maintenance effort. The most important concept in Germany is the SINFOS master data pool from CCG (Centrale for Coorganisation). The SINFOS data provide information for the goods receipt, the stockholding, the space optimization, the purchasing, and the verification of invoice and inventory (cf. CCG 1998).

However, even when such concepts are used, an enormous manual (re)processing effort remains in order to ensure consistent and current basic data. Thus, a major requirement for modern retail information systems is a flexible and efficient basic data management. SAP Retail provides various concepts here, such as reference articles, a comprehensive classification system and flexible search and copy functions. The following sections describe these in detail.

3.2.1 Classification System

Because of the large amount of master data, a central problem is finding the correct master data record. For example, a customer of a technical wholesaler that requires a replacement for a defective component will seldom be able to name the article number (of the retailing company). However, central functional or quality characteristics of the article are known. In such cases, an efficient retail information system must also permit the article search using such characteristics. A grouping is also useful when inadequate warehouse inventory means that a possible alternative article with similar characteristics should be sought that has not be defined as explicit substitute article. Similar situations are also conceivable when searching for customers and vendors.

The universal classification system that SAP Retail provides for these purposes allows a uniform structure and the same functionality to be used for all master data objects. Based on the object-oriented paradigm (cf., for example, Meyer 1997), class hierarchies can be built for the various master data types. The master data are assigned to the individual classes.

Class characteristics can be defined as mandatory or optional fields for each class. Possible characteristic forms can be specified as value ranges (10 mm-30 mm) or as discrete values (red, blue, yellow, green). The characteristics and the possible characteristic forms are transferred automatically to subordinate classes. (In the context of the object-oriented paradigm, this is termed inheritance.). At the level of the subordinate classes, it is possible to add additional characteristics and further restrict the value ranges for characteristics.

3.2.2 Vendors

The vendor master data groups all information concerning the vendors of a retailing company. The information can be differentiated whether it is generally valid, i.e., applies to the complete retailing company (or in SAP terminology: all clients), or whether it depends on the company code and possibly the purchasing organization or even the vendor sub-range. If the data depend on the company code or the purchasing organization, they can be defined differently for each company code or each purchasing organization.

The three central data storage levels (client – company code – purchasing organization) ensure that all data of a vendor can be maintained integrated in a logical vendor master record even for retailing groups with a complex structure. This can avoid the need to create several merchandise management system vendors for a physical vendor, with all resulting problems, which is sometimes observed in practice.

If such a differentiated distinction of the suppler data is not necessary, identical data can be used for all company codes and purchasing organizations.

The following list names the most important data of the three mentioned data storage levels of the vendor master data:

- *General data*

The general data include address data for the unique identification of the vendor (address, search terms), details about the communication paths (telephone, fax, data exchange and contact partner) and basic payment data (e.g., bank account details).

- *Company code data*

The company code data provide various details that are required by the billing and dunning programs or for reports to the financial authorities. These include details about withholding tax, interest payments, the automatic payment process, the dunning process, and correspondence-related details (responsible person at the creditor, account at the creditor, etc.).

- *Purchasing organization data*

The purchasing organization data cover all purchase-related details. In addition to general condition data (e.g., customer order currency, payment conditions, Incoterms, vendor-related minimum order value) and default values for articles (e.g., planned delivery time), these are mainly various purchase-related control indicators. The partner roles of the vendor are also defined at this level.

Partner roles

A vendor can assume various partner roles with regard to the retailing company. Thus, during the procurement process, a vendor can initially act as recipient of the order, then as goods vendor, and finally as invoicing party. The data stored in the vendor master data often depend on the vendor role (e.g., address or contact partner data). To represent this in the merchandise management system, various roles can be assigned to the vendor. It is possible that one or more of these roles reference different vendor master data records. The data stored there (which differ from the basic vendor record) are used automatically for the associated logistical and accounting functions.

The standard settings for SAP R/3 cover for example the following vendor roles: contract address, goods supplier, invoicing party (invoice presented by) and alternative payment recipient. The supplying plant (goods supplier) and the contact partner roles (contact person and contact person fresh produce) are in addition predefined in SAP Retail. Further roles can be defined during the customizing. Partner roles can be defined on different data retention levels (purchasing organisation, vendor sub-range and site level), so even complex partner role structures can easily be defined in SAP Retail.

It is also possible to specify in the customizing which roles are to be optional or mandatory and which must be unique, and so can be assigned just once per vendor. If roles are assigned more than once, the required role occupant must be selected in the associated subsequent function. For example, if two role occupants are defined as being *ordering address* for a vendor, the required order address must be selected when a purchase order is created. If, however, only one role occupant is defined as *ordering address*, this is done automatically.

The role concept also permits the assignment of task areas to the company's own employees or to the vendor's employees. As standard, a (general) contact partner and a perishables contact partner can be defined for each vendor in SAP Retail. A company employee who has responsibility for the vendor can be stored using the *responsible employee* partner role. Table 3-1 summarizes all standard vendor roles with their associated partner type.

Table 3-1: Predefined vendor partner roles in SAP Retail

Partner Roles (Purchasing Organisation Level)		
Abbreviation	**Partner Role**	**Partner Type**
VN	Vendor	Vendor
OA	Ordering address	Vendor
AZ	Alternative payment recipient	Vendor
PI	Invoice presented by	Vendor
GS	Goods supplier	Vendor
CA	Contract address	Vendor
CP	Contact person	Contact partner
IN	Contact person, fresh produce	Contact partner
ER	Employee responsible	Personnel master record
AL	Additionals	-
Partner Roles (Vendor Sub-Range)		
Abbreviation	**Partner Role**	**Partner Type**
OA	Ordering address	Vendor
PI	Invoice presented by	Vendor
GS	Goods supplier	Vendor
CA	Contract address	Vendor
CP	Contact person	Contact partner
ER	Employee responsible	Personnel master record
Partner Roles (Site Level)		
Abbreviation	**Partner Role**	**Partner Type**
OA	Ordering address	Vendor
PI	Invoice presented by	Vendor
GS	Goods supplier	Vendor
CA	Contract address	Vendor
CP	Contact person	Contact partner
ER	Employee responsible	Personnel master record

Because the partner roles can be defined at the level of the purchasing organization, different business partners can adopt various roles of a vendor in different purchasing organizations.

Figure 3-13 illustrates the use of partner roles using the example of a vendor that in the area of the southern purchasing organization of the retailing company has a different order address and also a different contact person from that in the area of the northern purchasing organization. However, the same contact person fresh produce is responsible for both areas. Note that roles not explicitly assigned to different business partners, such as the invoicing party role, are assumed by the base vendor (LF 32100 in this example).

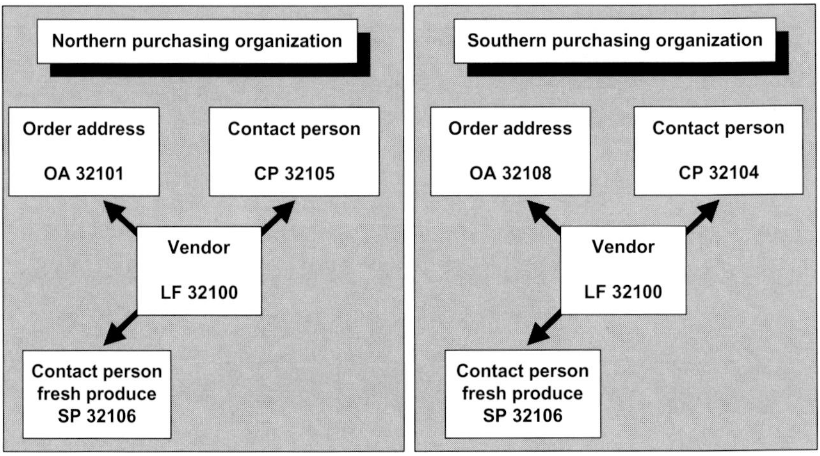

Figure 3-13: Sample partner roles

Vendor hierarchies

Since Release 4.0A, SAP Retail allows the use of special vendor roles to represent complex vendor hierarchies. Vendor hierarchies are used here to flexibly represent multi-stage hierarchical vendor structures. SAP Retail offers by default four-level vendor hierarchies.

Vendor hierarchies allow the adequate representation of sales groups, cooperative structures, complex group structures, retail outlet chains and generally hierarchically organized vendors with several locations. Vendor hierarchies are used in SAP Retail during the order processing to determine prices and to determine any end-of-period rebates (bonus calculation).

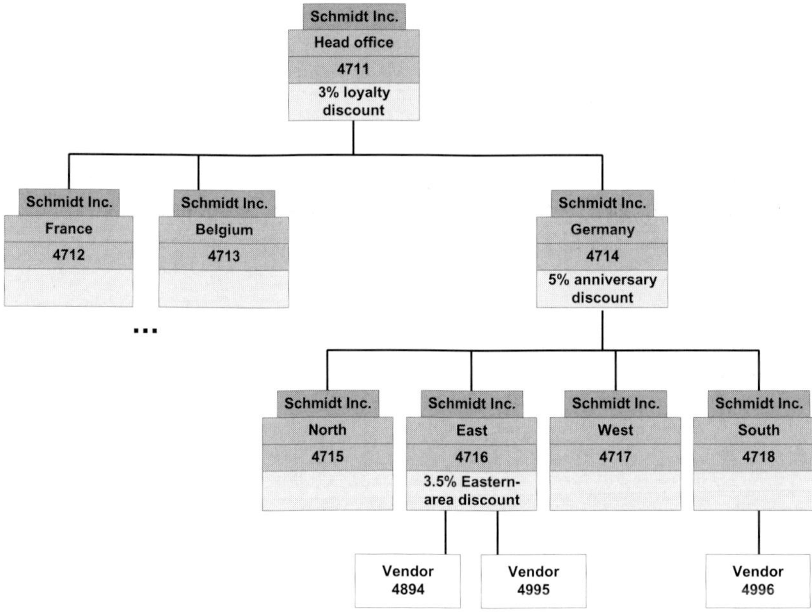

Figure 3-14: Use of vendor hierarchies to define the conditions of purchase

The use of an explicit representation of the vendor hierarchy can be shown easily with an example. An international group, Schmidt Inc., has factories and warehouses in Germany, Belgium and France. The individual subsidiaries and sales offices in Germany are structured regionally (north, south, west, east). A retailing company operating internationally now negotiates (additive) purchasing conditions at various levels.

- 3% long-term customer discount at the group level
- 5% anniversary discount for Germany
- 3.5% east discount for the eastern sales area

Without an explicit vendor hierarchy, the relevant conditions would need to be stored and updated redundantly for each individual vendor subsidiary. Figure 3-14 shows the definition of the purchasing conditions using a vendor hierarchy. The conditions are stored at the level for which they were agreed and so also apply automatically to all lower organizational levels of the vendor. Thus, in this example, the vendor 4996 receives a total discount of 8% (3% + 5%), whereas the vendors 4894 and 4895 receive a discount of 11.5% (3% + 5% + 3.5%).

Vendor hierarchies do not have a static character in the specified market conditions, rather they change dynamically as result of internal restructuring, company takeovers or company fusions. SAP Retail offers the capability to change nodes of the vendor hierarchy and to move them within the hierarchy. A complete time control ("valid from" – "valid to") permits advance changes to be maintained in the vendor hierarchy.

Purchasing info records

The purchasing info record is used in SAP Retail to represent the article-vendor relationship. A purchasing info record assigns an article to a vendor. If several articles can be obtained from a vendor, each has its own purchasing information record. Similarly, an article has several information records when it can be obtained from different vendors. This construct is used in purchasing to determine which vendors can deliver a specific article or which range of articles a vendor offers. The definition of the purchasing conditions, delivery conditions, etc., are handled using the purchasing information record, because these details are generally vendor-dependent and so cannot be maintained directly in the article master data.

Vendor sub-ranges

Vendor sub-ranges represent groupings of articles from the vendor's viewpoint. Because the vendor's grouping of the articles normally differs from the article grouping of the retail company, the merchandise categories cannot be used here. What is required is the representation of the vendor sub-range, in particular to permit the definition of conditions and contracts at this level.

Conditions that are valid for all articles of a sub-range can directly be defined at the level of the sub-range. Different conditions can apply to each vendor sub-range. When a vendor defines several sub-ranges, each article must be assigned uniquely to a vendor sub-range. This ensures that the conditions can be determined correctly. These are determined using an assignment of the vendor-specific purchasing information records to the individual sub-ranges. Figure 3-15 shows the basic structure.

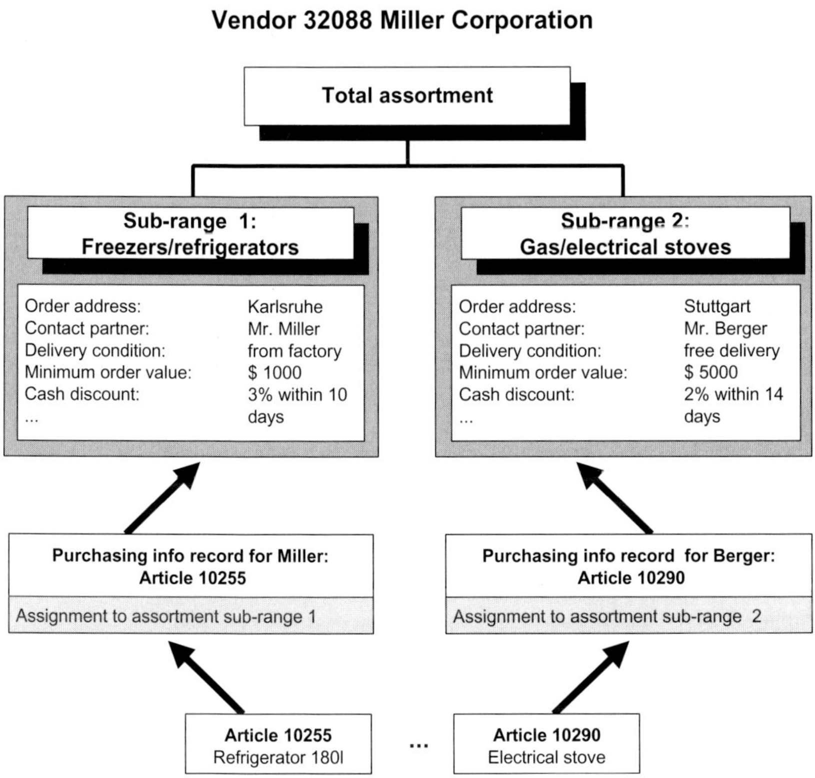

Figure 3-15: Vendor sub-ranges

Business partner roles can also depend on the vendor sub-range. Thus, it is conceivable that whereas a vendor ships small parts using a parcel delivery service (e.g., electronic components vendor sub-range), it delivers large appliances with its own trucks (e.g., household appliances vendor sub-range).

Default values can be defined at the vendor sub-range level for storing the article and purchasing information record, for example, planned delivery times, purchaser, purchasing cycle, delivery cycle (see Section 3.2.5)

3.2.3 Customers

A customer in the SAP system represents an organizational unit that receives goods. A differentiation is made between the customer categories: own companies, external customers, consumers and anonymous end consumers. Data different in content and extent must be stored appropriately for the customer category.

For example, whereas stock transfers between companies are possible, a stock transfer to an end consumer is not sensible.

Sites are automatically created as *internal customers* in SAP Retail. To ensure that the POS data can be assigned correctly, a dedicated *anonymous customer* should exist for each site. Extensive data can be stored in the customer master file for *external customers*, normally company customers. For example, several contact partners with the associated background information (e.g., visiting times, birthdays, etc.) can be maintained for each customer. In contrast, a *consumer* is used to represent an actual person (for example, end customers in the retailing business). Only a reduced customer master data are maintained for consumers. These data, which can be processed when the consumer is being updated in an input screen, contain the personal data (e.g., name, age, sex), address, bank account details and data for payment cards. Data at the level of the sales area and partner roles are not maintained for specific consumers, but are taken automatically from a defined reference customer. This allows a new consumer to be created with a minimum of data needing to be entered.

In addition to the various customer categories, *sales prospects* can also be managed in SAP Retail. Sales prospects represent potential customers whose data (for example, annual turnover, number of employees, industry segment or potential purchasing volumes) are needed for marketing purposes. The same basic functionality as for customers is available for the management and updating of the data for prospective customers; only irrelevant areas (e.g., billing) are deactivated. The change of a single attribute suffices to turn a prospective customer into a customer.

In addition to the classic customer master data, SAP Retail can also manage all contacts made with the customer (for example, customer visits, trade show contacts, provision of advertising material) to provide sales support. This contact history allows the complete communication with the customer to be followed in a clear manner.

The required information is passed from the customer master data to the operative business transactions. This allows all documents in sales to be based on the customer's relevant master data. This ensures that data do not need to be entered or updated redundantly.

3.2.4 Merchandise Categories

A merchandise category (sometimes also called material group) is a unique grouping of articles for the purpose of structuring and classifying of the complete assortment in depth and range. In Germany, the CCG has developed a reference merchandise category catalog which retail companies are increasingly using as basis for their merchandise category definition (cf. CCG 1997). The use of consistent merchandise categories simplifies cross-company analyses and comparisons.

Merchandise category hierarchies that allow a merchandise category to be split into further sub-merchandise categories are usual in the retail sector (see Figure 3-16). The CCG merchandise category catalog is also built hierarchically. For example, the CCG merchandise category "2400 ladies' woolen coats" is assigned to the higher level merchandise category "240 ladies' coats and jackets", which itself is a part of the merchandise category "24 ladies' apparel".

In contrast to standard R/3, SAP Retail supports merchandise category hierarchies and permits a practically unlimited number of hierarchy levels. The merchandise category hierarchy is represented using the classification system of SAP Retail (see Section 3.2.1). The individual merchandise categories and merchandise category hierarchy levels each corresponds to classes to which the articles are assigned uniquely.

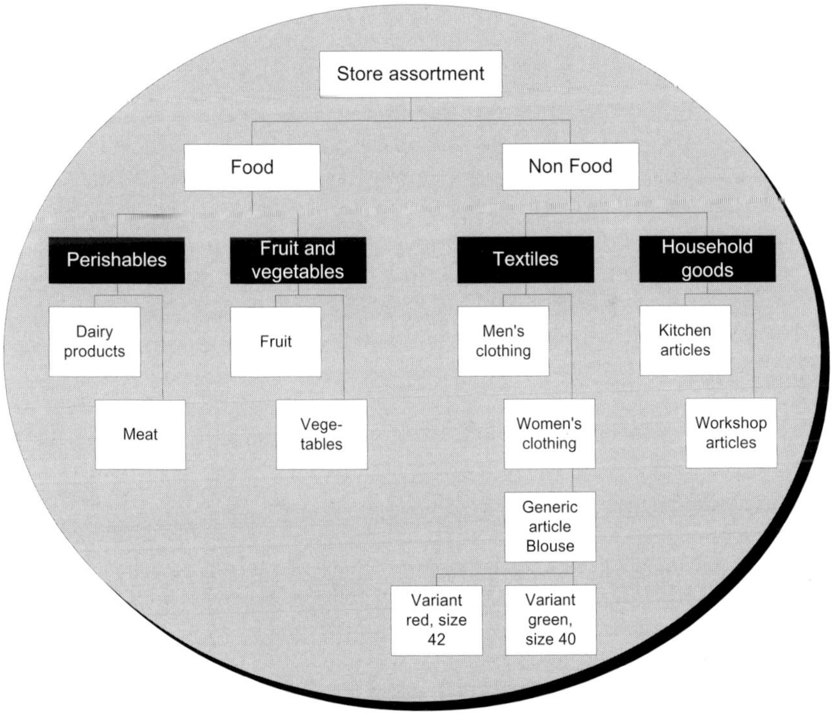

Figure 3-16: Merchandise category hierarchy

Merchandise categories can be used in SAP Retail for various purposes:

- The central planning, and analyses and reports for the controlling are normally performed at the merchandise category level. The merchandise category hierarchy offers here the advantage that the use of various hierarchy levels allows analyses to be performed at different aggregation levels.

- Conditions (in particular sales conditions) can be stored at the merchandise category level. This avoids the need to maintain similar conditions for each individual article redundantly.
- To simplify the storing of articles, suggested values can be defined at the merchandise category level using a so-called *merchandise category reference article*. The reference article cannot be used operatively in the materials management processes, but serves only to store the suggested values that are invoked automatically when a new article is created in the associated merchandise category. For homogeneous merchandise categories, this concept allows a significant reduction of the data entry effort when new articles are stored.
- A *merchandise category article* can be defined for each merchandise category. This value-only article can be used to register the values for sales that cannot be assigned to an exact article. Specific articles contained in the merchandise category can be removed from the merchandise category article to allow them to remain as precise articles. This capability is used in food retailing, e.g., for exotic fruits that are often managed and sold as individual articles even though they belong to the "Fruit" merchandise category, for which normally inventory management is not made at the single article level.
- The *cross-merchandise-category value-only article* is specially designed to satisfy the requirements of the retail sector. This is required when the cash registers at the POS have fewer merchandise category keys than the number of defined merchandise categories. Cross-merchandise-category value-only articles can be assigned to any node of the merchandise category hierarchy.

Reclassification of articles

The merchandise category hierarchy and the assignment of articles to merchandise categories are often subject to changes over the course of time. Although the simple reassignment of a hierarchy node or the change of the merchandise category assignment of an article does not initially represent a technical problem from the DP viewpoint, ambiguities and interpretation problems can result. The changes of the merchandise category hierarchy remove the ability to make comparisons with the previous year's value in the controlling. Unless appropriate corrective measures are adopted, this threatens to produce inconsistent comparisons. Problems also arise for all definitions made at the merchandise category level (merchandise category article, merchandise-category-specific article attributes, merchandise category conditions, merchandise category contracts, etc.). What, for example, needs to be done for conditions that were defined at the merchandise category level when an article is removed from this merchandise category?

Because of these problems, merchandise category changes in practice should be reduced as far as possible. Sales promotions or other short-term "groupings" of articles should not result in changes to the basic merchandise category hierarchy. In addition, "theme-based" groupings of articles associated with ECR or category management concepts – possibly short-term – because of their dynamics, are sometime better represented outside the merchandise category hierarchy. Sometimes, the parallel use of several equivalent article hierarchies (so-called multiple article hierarchies) is also suggested for such applications. The high complexity means the alternative constructions for article grouping appear to be more appropriate, in particular for analysis purposes.

SAP Retail does not provide the parallel use of equivalent article hierarchies. Although a reclassification procedure is offered to change the merchandise category hierarchy and the assignment of articles to merchandise categories, it currently exhibits a number of restrictions: Statistical changes are not initiated automatically, the historical document data remain unchanged, all hierarchy conditions are retained and may need to be changed manually.

3.2.5 Articles

The article is the central object in almost all functional areas of a merchandise management system. Various article-related details integrated in the article master data are needed to control these procedures. To structure the article information, which may include several hundred single attributes for each article, the article master data are subdivided into various thematic subareas (so-called views). Because the views apply to different organizational levels, they can be valid in different areas.

SAP Retail contains definitions for the following views in the article master data:

- *Basic data*
  The basic data cover the identifying attributes (article number, UPCs or EANs), the descriptive attributes of the article (basic text, references to picture files, and, possibly, sound files), the internal article grouping (merchandise category assignment) and various units of measure. The basic data are valid throughout the company.

- *Listing data*
  The listing data, which have particular importance in chain store companies, cover the listing period, the listing procedure and, possibly, the fixed assignments of articles to assortment modules and assortments.

- *Purchasing data*
  The purchasing data cover vendor-dependent details for the deliverability of the articles, for the vendor sub-range, for minimum quantities and conditions, and for the planned delivery times. The few vendor independent contracting data include a contracting or contracting group assignment that defines the contracting-organizational responsibility for an article.

- *Logistics data (distribution center)*
  In SAP Retail, the logistics data (DC) cover not only the central logistics data, such as storage location, storage bin, picking area and loading group, but also the purchasing and forecasting parameters. Various model parameters can be defined depending on the selected model. The contracting assignment also takes place at this level. The logistics data (DC) can be defined differently for each distribution center.

- *Logistics data (store)*
  The logistics data (store) largely correspond to those of the distribution center and can also be maintained separately for each store.

- *Sales data*
  The sales data cover the sales-specific data, in particular the article selling prices. They are defined at the level of the distribution chain and, to some extent, at the individual store level.

- *POS data*
  The POS data consist of the various cash register receipt text items (multilingual cash register receipt texts can be stored for each issuing unit) and control parameters for the POS upload and download (see Section 4.7). With the exception of cash register receipt data, which are defined at the client level, the POS data can be defined separately for each distribution chain.

The provided and needed article characteristics differ widely in the individual retailing industries. Whereas country of origin, caliber sizes and grades are significant article characteristics in the perishables area, size details and color variants dominate in the textiles area. To reduce the input and updating effort for the article data, ergonomically designed article master data entry screens that can be customized to these specific requirements are a significant prerequisite.

In SAP Retail, these definitions are mainly made using the *article type*. As part of the customizing, the settings for the screen sequence and the field definitions of the individual entry screens can be defined for each article type. Standard settings are provided for the following article types: food, perishables, beverages, non-food / hard goods, fashion and textiles. These settings can be freely extended.

Furthermore, the creation of articles is simplified through the provision of implicit default values (e.g., based on the merchandise category reference article or the vendor sub-range details) or with the explicit specification of a freely selectable reference article. However, the article updating represents a central problem in retailing that should not be underestimated. In particular, the possibility to make a differentiated article listing and a possible detailed definition of the logistics and sales data at the store or distribution center level for a large number of stores can result in an enormous input and maintenance effort. As possible solution, SAP Retail allows a sample store to be defined for each distribution chain. The data defined for the sample store are then used as template for all those stores that are assigned to the distribution chain. Updating programs are available to perform changes to terms and conditions or selling prices for many articles or stores. For example, selling prices can be defined or changed not only specifically for each store, but also for a store group.

Article variants

Articles that differ only in a few characteristics exist in almost all retailing areas. Examples are different taste variations for chocolate bars or yogurt, different colors for light switches or different sizes for textils and shoes. Neither the grouping of all variants of an article nor the creation of a separate article for each variant is sensible. The first is not appropriate because it does not provide any useful information about the individual variants. Thus, the inventory information that 70 pairs of jeans in unknown sizes and colors are on stock is not particularly useful for purchasing. The representation of the variants as separate individual articles is also not appropriate, in particular, because of the updating effort and performance aspects.

The generic article construction is a possible compromise between the two previously mentioned concepts. The generic article represents a logical base article for which variants can be created. The identical data used for several variants and the possible different forms of the variant characteristics are maintained in the generic article. When variants are created, only the required variant characteristic forms or the required combinations of variant characteristic forms for multiple variant characteristics need to be selected. An article master data record is created and automatically filled with the data from the generic article for each article variant. Any required changes, such as different prices for large sizes, can be made subsequently in the affected sizes. A special *variant matrix* can be used in SAP Retail to simplify the representation of the variants of a generic article (e.g., in the purchasing or in the goods receipt).

Structured articles

Structured articles are those that consist of several individual articles. Depending on the type of the individual components and the purpose for grouping, *displays*, *sales sets* and a *prepacks* can be differentiated (Figure 3-17).

The central goal in the use of structured articles is the reduction of the input and maintenance effort, and the logistics effort. Sales enhancement and product presentation aspects also play a decisive role for the sales set and the display.

Figure 3-17: Structured articles in SAP Retail

A *sales set* represents a grouping of articles from the sales viewpoint with the goal of sales enhancement or product presentation and for which a special set selling price applies. The special feature of a sales set is that it is formed from individual articles often just at the retail company. The retail company buys individual articles and sells these exclusively or additionally as set. Examples for sales sets are gift baskets, porcelain services or the selling of grouped kitchen cabinets.

A *display* is a unit built for the purpose of product presentation. The contained articles can come from a single assortment or can be a mixture (for example, different types of chocolate). A display is often produced by the vendor. Displays are normally separated into their individual articles only in the store. Then, in contrast to the sales sets, these articles are only sold individually. Thus, there is no sales-side “display price” for the end customer.

A *prepack* is a grouping of articles primarily for logistical reasons. The contained articles must be different variants of a generic article (for example, different sizes of a ladies’ shoe). The other properties correspond to those of the display.

Article number

For identification purposes, a unique article number is assigned to every article. In practice, a differentiation can be made between two forms of article numbers: purely identifying article numbers and classifying article numbers. Because classifying article numbers can directly provide certain article properties, these are also designated as descriptive article numbers. Frequently, the vendor or the merchandise category is included in descriptive article numbers, more seldom an indicator used to distinguish between third-party articles and warehouse articles.

SAP Retail supports both externally assigned descriptive article numbers and internally generated consecutive article numbers. An automatic partial generation of article numbers can also be used, for example, after the manual specification of the first three digits to identify the vendor with the next free number within the vendor being appended automatically.

EAN / UPC

From the wide range of article codes, the European Article Number (EAN) (cf. CCG 1995, pages 79-91) and the American UPC (Universal Product Code) represent the most important article coding schemes. The first is the dominating article code in Europe, the latter is widely used in North America.

The goal of both concepts is to use a central number assignment to avoid a overlapping of numbers and thus to realize uniqueness of article numbers. Other than the fact that the UPC is only a 12-digit code and uses a slightly different coding (see Figure 3-18), it is quite similar to the EAN. Thus, the EAN concepts described in following the section and all referencec to EAN also apply to UPC.

Every logistical unit of an article (e.g., carton, enclosing carton, pallet) normally possesses a 13-digit EAN that permits a unique identification of the article and the logistical unit. The manufacturer prints the articles with the associated EAN (*manufacturer EAN*). Thus, the EAN is available in all subsequent logistical and sales processes.

Retail companies can also produce their own EANs, for example, to identify articles that have not been numbered by the manufacturer. Such retailer-internal EANs, which, because of the missing protection against overlapping assignments, cannot be used between sites, are designated as *instore EANs*. Another special case results for packed articles that vary in weight and quantity (e.g., fruit, cold meat, cheese) for which special *weight/price EANs* are used to provide a direct reading of the weight or the price at scanner cash registers.

Figure 3-18 shows the structure of the scanable EAN-13 and the UPC barcode.

Figure 3-18: The EAN-13 and UPC as barcode

SAP Retail supports all mentioned EAN (and the corresponding UPC) types. Because a freely-definable test algorithm can be assigned to each number type, arbitrary user numbering schemes (maximum 18 digits) can also be defined and used analog to the standardized EAN types. An EAN can be stored for each unit of measure for an article (item, carton, pallet).

Packaging units

Procurement, warehousing and distribution of the articles are performed not only in one unit of measure. Articles are assigned one or more packaging units (synonym: units of measure, logistical units). Typical logistical units are the pallet, the enclosing carton and layers. They can be split into smaller subunits in several steps (pallet, layers, 6-pack cartons, individual items).

SAP Retail permits a flexible definition of arbitrary logistical units. A conversion between different units of an article is possible using an internal base unit. Use purposes can be assigned to the individual logistical units. This purpose can, for example, indicate whether a unit is a purchase order quantity unit, a delivery/ issuing unit or a sales unit. If several units are defined for certain functions, the required unit of measure has to be selected in the associated business transaction.

The units of an article, which by default are defined in the article master data, can be overwritten at the vendor level (purchase order quantity unit) or at the level of the distribution chain (delivery/issuing unit and sales unit).

Article data model

The SAP data model allows a good representation of the previously described dependencies of the article on other objects and the various article types (see Figure 3-19).

An article must be assigned to an article type and a merchandise category (referential relationship). Because the article type has such a fundamental importance (e.g., for the control of the screen display sequence and for the definition of article master data fields), an article can be created only after an article type has been specified.

The retail article represents a specialization of the general article/material and can itself be further divided into article types (individual article, the various structured articles and value-only article). The model also shows that any number of generic article variants can be created for a generic article.

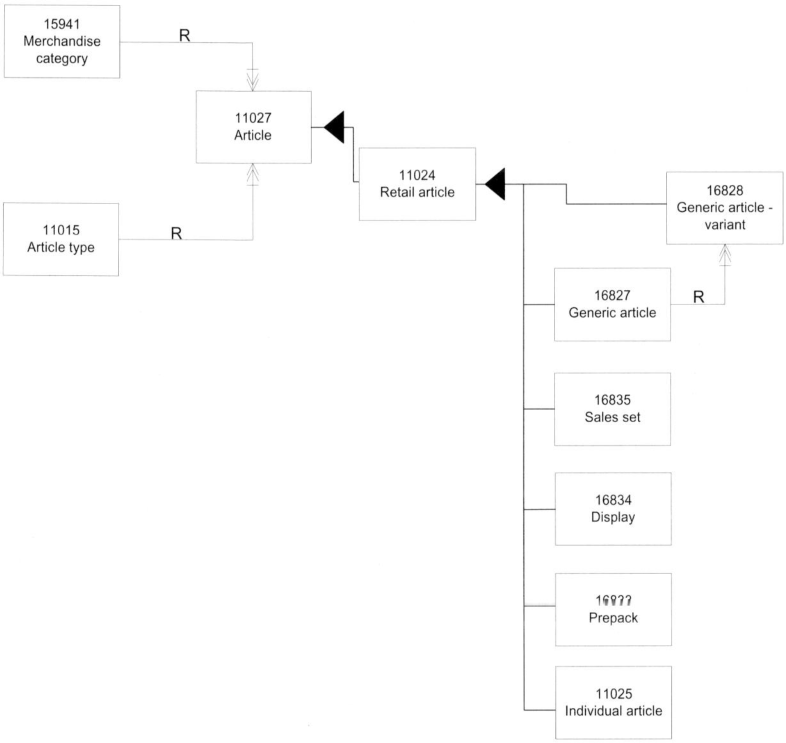

Figure 3-19: Article data model (selection)

3.2.6 Conditions

The condition management covers the agreement and management of delivery and payment conditions, prices and discounts. Both the purchase-side and the sales-side condition management are characterized by an extreme range of conditions (see Figure 3-20).

In the retail sector, the conditions (in particular, discounts) are agreed bilaterally in so-called annual negotiations between individual manufacturers and retail companies. The enormous range of purchase-side discounts, in particular in the retail sector, results from the purchasing power of large retailing groupings that hinders a reduction of previously agreed discounts and discount rates. However, complex condition structures from the manufacturers, also by retailers in the wholesale area, can be used to make it more difficult for the retail company or the wholesaler customer to make comparisons with the delivery prices from alternative supply sources.

Condition Types Retail company A	
1. Special condition 2. Base quantity discount 3. Immediate discount 4. Bonus for direct purchase 5. Merchandise category discount 6. Bonus for warehouse procurement 7. Incremental remuneration for total turnover 8. Central remuneration 9. Quota remuneration 10. Assortment coordination remuneration 11. Annual bonus 12. Low complaint rate remuneration 13. Factoring discount balancing 14. Cash payment discount 15. Extended payment date 16. Free sample 17. Price increase minimum notice twelve weeks	18. Promotion conditions (runtime six weeks) 19. Pallet advertising cost allowance 20. Damages for passing on of information to third-parties one percent of annual gross turnover 21. Opening conditions depending on the size of the market, in DM absolute 22. Reopening condition 23. Value date 24. Remuneration for product support service 25. Remuneration for price labeling and shelf maintenance 26. Advertising remuneration - without insertion - with insertion 27. Distribution discount 28. Listing discount / advertising cost allowance 29. Article immediate discount

Source: Steffenhagen (1995), p. 39 f.

Figure 3-20: Sample conditions of purchase for a retail company

The ability to represent the previously mentioned conditions in a merchandise management system requires a structural differentiation of the conditions. Conditions can be basically differentiated whether they concern invoicing conditions or subsequent conditions. When a differentiation is also made for conditions that depend on one or more order transactions, and those that apply generally, three condition types can be identified (cf. Becker/Schütte 1996, p. 154):

- **Basic conditions**: Conditions that apply generally and thus independent of the individual business transactions (e.g., general reseller discounts).
- **Business-transaction-dependent conditions**: Conditions that depend on one or more conditions associated with a business transaction (for example, quantity scale discounts).
- **Subsequent conditions**: Conditions applied subsequently after a certain time (and thus designated as *subsequent conditions, subsequent rebates* or *volume rebate stipulations*).

Whereas the first two condition types are clearly defined at the time of the business transaction and thus can be monitored during the invoice verification, the subsequent conditions are unreliable conditions that are determined only ex post (at the end of the settlement period) whether and at which rate they apply.

However, because of the importance of subsequent conditions in the contracting, a permanent tracking of the previously attained subsequent condition claims is necessary. Should there be signs that the stipulated volume rebate will not be achieved, this allows measures to be adopted early (e.g., procurement of articles as part of promotions to increase article sales or make investment purchases to reach the agreed buying volume despite lagging demand).

SAP Retail provides almost identical updating instruments and constructions for purchasing-side and sales-side pricing conditions. The three central tasks in both cases are the updating of the (standard) prices, the specification of surcharges and discounts, and the definition of subsequent rebates. Conditions can be defined flexibly for a wide range of data storage levels (on the buying-side, for example, vendor, vendor sub-range, article; on the sales-side, for example, customer, merchandise category, article). Because often different conditions for the same article apply depending on the vendor, article-related conditions of purchase are not stored directly in the article master data, but in *purchasing info records* (see Section 3.2.2).

SAP Retail supports all usual condition types, such as value scales, quantity scales, absolute and percentage surcharges and discounts, and free-goods discounts (inclusive and exclusive). Because all conditions can be defined with a time reference ("valid from... to"), the advance entry of future prices and conditions is possible.

The pricing is made using a free-definable condition scheme that specifies which price applies to the article, which surcharges and discounts are used, and in which order these are applied (also refer to the example for sales-side pricing on page 87). The possibly complex calculations can be represented clearly – with all intermediate results – in the condition screen.

The *pricing* in SAP Retail is tailored to the requirements of the retailing industry. This special form of pricing permits a direct calculation of the selling price based on the purchase price and a defined markup rate and other parameters (e.g., automatic price point rounding). Thus, in contrast to "normal" sales-side pricing, no base selling price needs to be specified, but rather a direct reference can be made to the purchase price. It is also possible to calculate parallel prices for store chains with different prices or different customer groups (in the wholesaling area).

The pricing can be used as a *single-step* or a *two-step price calculation*. For the single-step price calculation, the retail price for the store is determined directly based on the purchase price. For the two-step price calculation, a distribution center price (central warehouse price) is first calculated based on the vendor price and, in a second step, the retail price and possibly a wholesaler customer price derived from the distribution center price.

Figure 3-21 shows a comparison of the three calculation methods.

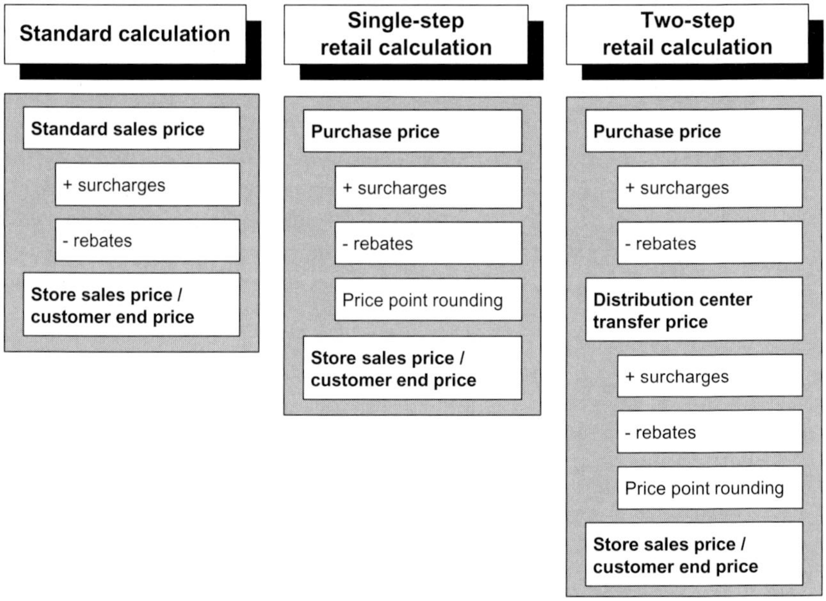

Figure 3-21: Methods for price calculation in SAP

3.3 Procurement

The procurement process covers the subareas: contracting, purchasing, goods receipt, invoice verification and creditor accounting. The contracting handles the administrative tasks (e.g., administration and updating of the master data) that form the basis for the procurement logistics. The procurement logistics has as task the requirements-related and economical supply of goods to the retail company. The current warehouse inventory and the forecast future article demand are used to determine the purchasing quantities and to generate purchase orders. The manual or automatic invoice verification and the creditor accounting complete the procurement process. Because legal restrictions greatly standardize the accounting procedures and functions, which accordingly are covered comprehensively by various financial accounting programs, including the FI module of the SAP system, the following sections do not further discuss the creditor accounting.

When we consider the various arrangements of the procurement processes with regard to the information and logistics flows, three fundamental procurement types can be differentiated (see Figure 3-22; also refer to Section 2.6 for third-party business):

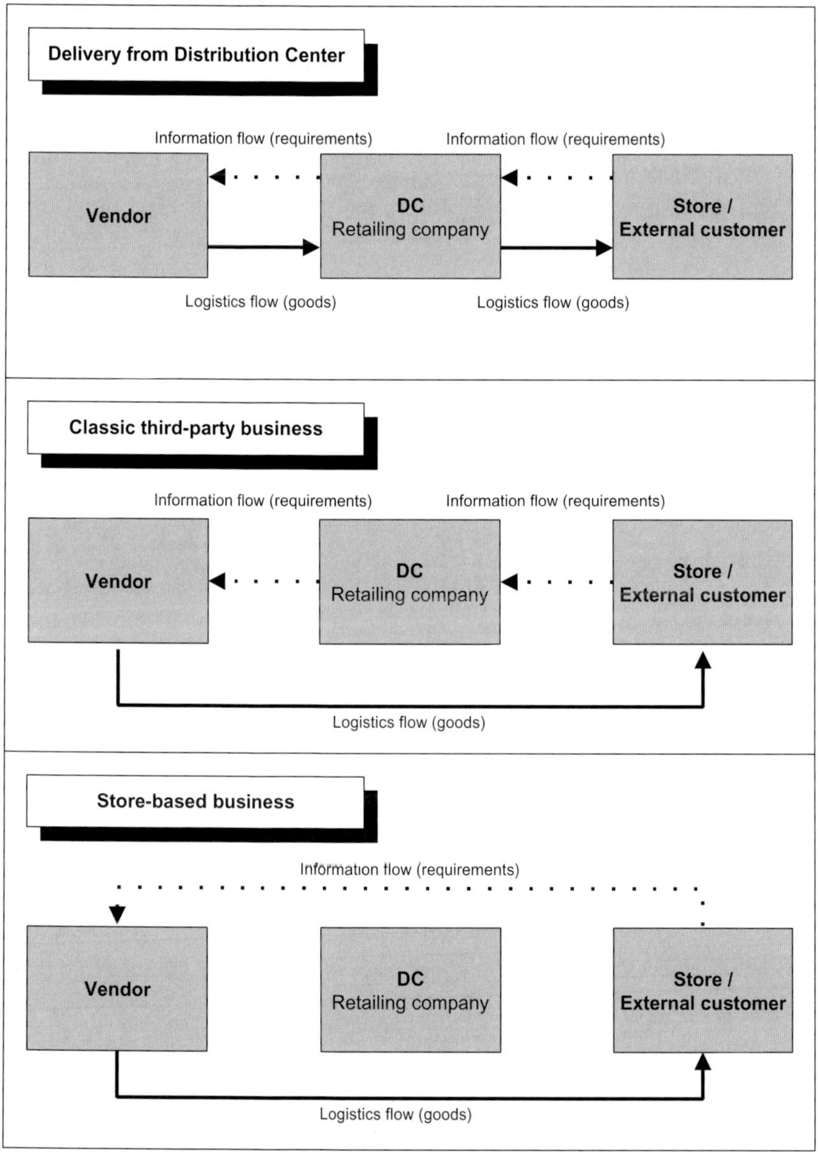

Figure 3-22: Basic forms of procurement

Delivery from distribution center

In case of a delivery via a distribution center (DC), the retailing company adopts its classic assortment forming and demand bundling function. The recipient (the customer or the store) no longer has direct contact with the vendor. With a single purchase order to the retailing company, the recipient can order goods from different vendors (assortment forming function). On the one hand, the distribution center can bundle various recipient demands and forward these together to the vendor (demand bundling function). Because the distribution center assumes a stockholding function, the ordering with the vendor is decoupled from the recipient ordering with the retailing company. This means that the distribution center orders goods for stock, stores them and delivers them for the corresponding purchase orders from the recipients.

Although this procedure is certainly efficient for smaller and irregular requirements from the recipients, the stock placement and stock removal costs become significant for larger (and regular) deliveries to the recipients. Newer logistical concepts, such as *cross-docking* and *flow-through* avoid the mentioned stock placement and stock removal costs (cf. Hadjiconstantinou 1998). For cross-docking, the recipient's purchase orders are forwarded directly to the vendors. The vendors deliver to the distribution center the goods already packed for the recipient. For example, if two recipients each order half a pallet of the same article, the vender delivers two half pallets. At the goods receipt, the individual pallets are assigned to the customers and transported directly to the goods shipping. This avoids both the stock placement and the stock removal of the goods.

The procedure for flow-through is basically identical to cross-docking, however, the delivered pallets may need to be broken up into smaller units of measure in the goods receipt. Thus, the vendor in the above example would not deliver two half pallets, but one complete pallet, which then must be divided in the goods receipt area of the distribution center. Consequently, the cross-docking and flow-through differ in who performs the goods picking. Whereas for cross-docking, this task is passed completely to the vendor, flow-through normally requires a post-processing in the goods receipt area.

Classic third-party business

For classic third-party business, the distribution center collects the recipient demand and forwards it bundled to the vendors. The purchase order lists the separate requirements of the individual recipients. This allows the vendor to directly supply the recipients. The central third-party business is used, in particular, in centrally organized retail companies in which the stores sometimes no longer report their requirements, but rather these are determined at the head office.

Store-based third-party business

A direct contact between the recipient and the vendor takes place in store-based third-party business. The recipient passes its requirements to the vendor and the vendor directly supplies the recipient. The retail company's distribution center does not perform any logistical functions. However, the head office of the retail company is included in the invoicing and payment process.

3.3.1 Contracting

In addition to the administration of the contracting-related master data (e.g., articles, vendors and conditions of purchase) that have already been discussed in detail in Section 3.2, the contract administration and the vendor evaluation (as subfunction of the vendor administration) are assigned to the contracting.

Contract administration

A *contract* is an agreement of a retail company with a vendor with regard to the article that is to be purchased within a specific time period. Contracts can be defined on the basis of article quantities (*quantity contract*) or on the basis of article values (*value contract*).

Scheduling agreements represent a further tangible form. In contrast to the value and quantity contracts, a scheduling agreement covers concrete partial delivery quantities and fixed delivery dates (so-called schedule lines). Scheduling agreements are particularly suitable for articles with requirement quantities that can be easily forecast.

SAP Retail supports contracts and scheduling agreements both at the single article level and at the merchandise category level. It is possible to define several overlapping contracts for an article that run parallel over time. The purchase orders then must specify which contract is referenced.

Vendor evaluation

The vendor evaluation is a time-period-related assessment of the vendors that is primarily useful for the contracting department for the annual negotiations (Harting 1989). For an exact vendor assessment, data from the individual business transactions, i.e., form each individual order and each goods receipt item, must be received and evaluated. An integrated retail information system can automate this operation. Significant evaluation criteria used in practice are price (more correctly: the price level), the goods quality and the delivery quality. The *price level* can be determined using comparisons with alternative vendors. The *delivery quality* results primarily from the delivery reliability. These can be derived from the planning data and the actual goods receipt quantities and dates. The goods quality can be determined using the goods receiving inspection, possibly supplemented with evaluations of customer complaints and customer audits.

SAP Retail provides a flexibly-structured vendor evaluation system. As standard, the main criteria provided for the vendor evaluation are price, quality, delivery and service. In additional 95 main criteria, which can also be formed from various subcriteria, can be freely defined. It can be specified for the individual criteria whether the assessment is to be entered manually or should be determined automatically by the SAP system (through the assessment of the operative business transactions). Various assessment functions and graphical displays permit a comprehensive analysis of the evaluation results.

3.3.2 Purchasing

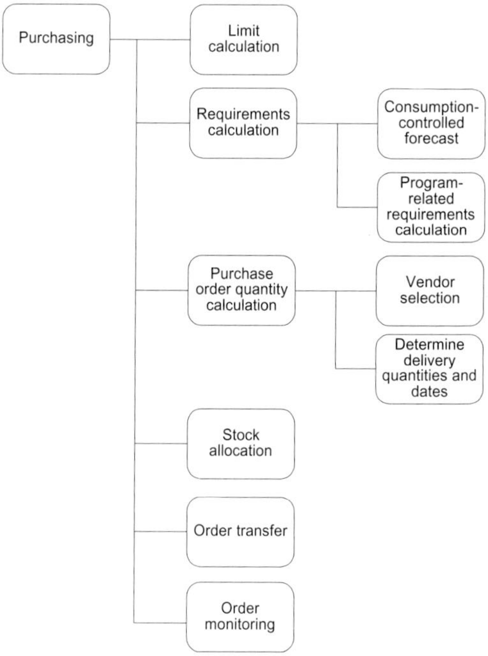

Figure 3-23: Purchasing subfunctions

The purchasing area covers the more mid-term tasks of limit calculation, requirements calculation, the purchase order quantity calculation, the operative order execution and the order monitoring. The (purchasing-oriented) tasks associated with the stock allocation are also assigned to the purchasing area (see Figure 3-23).

Limit calculation

The limit calculation is a control instrument for the procurement of goods that is used with the aims of cost reduction and safeguarding liquidity. As part of the limit calculation, procurement upper limits are typically defined for the merchandise categories to ensure that the procured goods quantities correspond to the sales planning. The limit calculation restricts the retail stock planners' scope.

SAP Retail permits the definition of the procurement limits at the level of articles, merchandise categories, vendors and retail stock planners, etc. The limit definition can be based for the purchase price or the selling price. In addition to the usual value-based definition of limits, quantity-based limits (for example, maximum 1000 units of an article) can be used. The quantity-based limit planning is particularly suitable for articles with highly fluctuating prices (e.g., in raw materials trading).

The limit monitoring is performed either directly as part of the order entry or periodically in the form of an exception reporting. The first-mentioned method has the disadvantage that the complex checking routines degrade the performance of the SAP system.

Requirements calculation

The requirements calculation has the task of determining the future article requirements. This can be done either by grouping available concrete sales orders (*deterministic requirements calculation*) or by forecasting future article requirements based on sales quantities from the past (*stochastic requirements calculation*). Whereas the deterministic requirements calculation is normally used in industry, it is used seldom in retailing. One use is in promotion business in which the promotion order quantities for the customers and stores are provided with such a large lead time that their sum produces the purchase order quantity of the distribution center.

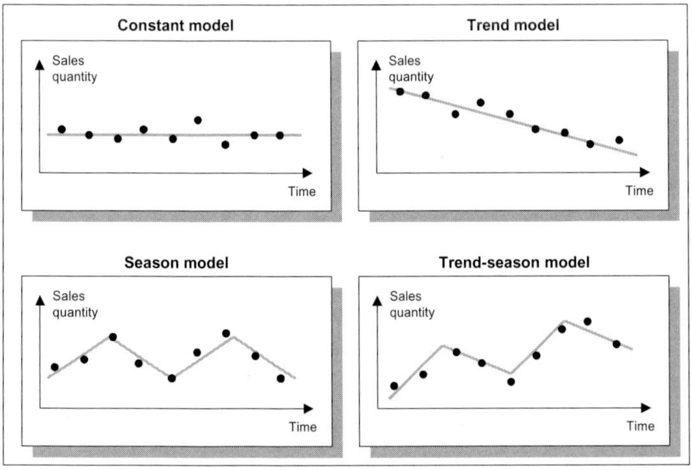

Figure 3-24: Forecasting models

The stochastic requirements calculation dominating in retailing forecasts the future article requirements using mathematical methods, such as exponential smoothing of the first, second or third order. The choice of the correct demand model is decisive for the quality of the forecast results. A differentiation is made between four models: the constant model, the trend model, the season model and the trend-season model (see Figure 3-24).

The *constant model* assumes a constant demand that is subject only to random fluctuations. In contrast, the *trend model* assumes a long-term demand trend – positive or negative. The *season model* is based on the assumption that the demand fluctuates cyclically (e.g., during the year). The base trend for the *trend-season model* is subjected to a cyclic demand trend. SAP Retail allows the use of all discussed base models. The specification of the model type and the forecast parameters can be made individually for each article.

Various special systems designed for an optimized calculation of article requirements (e.g., E3TRIM from E3 Germany GmbH or LOGOMATE from Remira IT GmbH), which can be coupled with SAP Retail using interfaces, exist to determine the requirement quantities.

Purchase order quantity calculation

During the purchase order quantity calculation, the previously determined requirements data are used to specify when and in which quantities articles are to be obtained from which vendors. The purchasing in retailing is normally performed vendor-related. All required articles are ordered on the vendor's day of ordering, which is calculated from its delivery day and the delivery time. The purchase orders can be created in various ways that can be differentiated according to the degree of automation:

- *Manual purchasing*
  In manual purchasing, the retail stock planners determine the purchase order quantities. Various data from the merchandise management system, such as past sales, warehouse inventory, packaging units and minimum order quantities, can serve as basis. This procedure is sensible only for articles required infrequently (or one-time articles) and for articles for which it would be difficult to determine the demand model. For articles seldom planned, the question should generally be asked whether the updating effort for the purchasing parameters for an automatic purchasing is not too high compared with the benefits.

- *Automatic order recommendations with manual order release*
  In this semi-automatic purchasing procedure, the merchandise management system automatically generates order recommendations. The forecasted demands, warehouse inventory and a wide range of optimization parameters (packaging units, minimum order quantities, scale prices and discounts, target stocks, vendor delivery cycle, etc.) serve as basis. However, the purchase order is not automatically sent to the vendor, but must be released explicitly by the retail stock planner.
- *Automatic order recommendations and order release (automatic purchasing)*
  For this fully automated form of the purchasing, the order recommendation and the order release/transfer are produced automatically. This procedure is particularly suitable for articles with a high turn-over rate having a relatively constant demand trend.

SAP Retail provides various default processes for manual, semi-automatic and automatic purchasing. For the semi-automatic purchasing, the purchase order quantities can be changed or further order items can be added prior to the manual release as part of the so-called *additional planning*. SAP Retail provides the reorder point planning, the time-based purchasing and the replenishment planning for system-supported purchasing procedures (for the semi-automatic and automatic purchasing). SAP Retail supports an automatic purchase order quantity rounding (such as for quantity optimization to complete logistical units of measure) for these purchasing procedures.

For *reorder point planning*, a replenishment level is manually defined for each article or automatically calculated by the system on the basis of the expected article demand and the replenishment lead time. When the available stock falls below the replenishment level, an order recommendation is generated automatically.

For the *time-based purchasing*, a periodic purchasing tag is specified for each vendor (less often, per vendor sub-range). An order recommendation is created automatically on the purchasing day for all pending article demands for the vendor (or vendor sub-ranges).

For the *replenishment planning*, the required "replenishment quantities" (normally for stores) are derived directly from the sales data of the POS systems ("one sell – one buy"). Figure 3-25 shows the schematic procedure of a replenishment disposition in SAP Retail as an event-controlled process chain (EPC) (cf. Scheer 2000, p. 125 f. for the EPC notation).

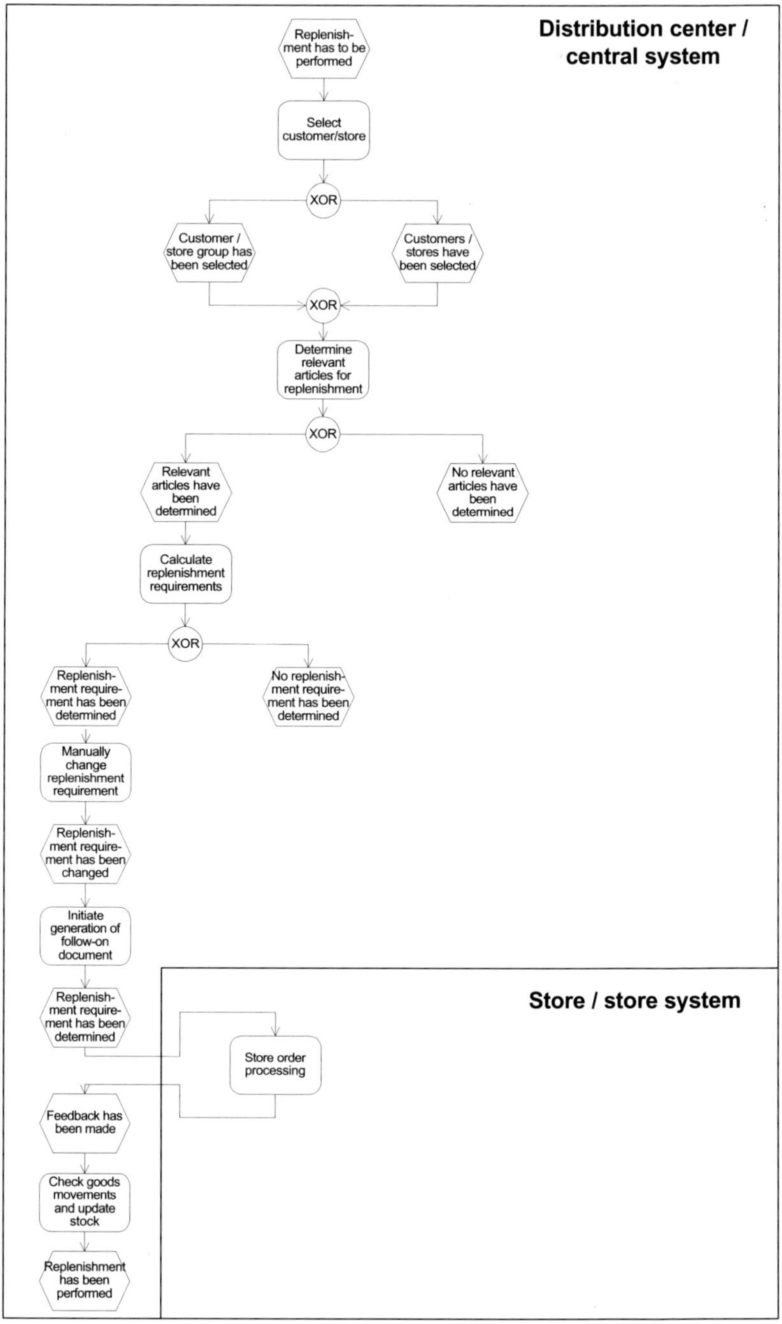

Figure 3-25: Replenishment process model (simplified)

In SAP Retail, the purchasing procedure and the associated purchasing parameters can be specified individually for each article or at a higher level (e.g., merchandise category).

Procedures for *order bundling*, both in store-based retailing (bundling of the individual store orders) and in warehousing (bundling of parallel pending sales order-related purchase orders and planned warehouse orders), can be used with success to create optimized procurement quantities. The order bundling concentrates both on a minimization of the transport costs through the optimum utilization of the transport means and on achieving more favorable purchase prices by making use of scaled conditions. As part of the order bundling, SAP Retail allows the grouping of purchase orders.

In addition to the previously described classic purchasing procedures, SAP Retail also provides the required functionality to support *investment-buying*. Investment-buying is a method used to determine article requirements while taking account of future cost changes (e.g., announced purchase price increases). The aim of the investment buying is to purchase suitable articles in appropriate quantities, on time, as cheaply as possible. To determine the procurement quantity, the system compares the current stocks, the forecasted requirements, and the order, transport and warehousing costs in a calculation based on the return-on-investment (ROI). This determines whether (and at which date and in which quantity) it is economic to procure articles even before their actual requirement date. Such a procedure can be adopted, for example, to avoid a price increase announced by the vendor or to achieve higher scaled levels. Provided the calculated benefits exceed risk limitations parameterized in SAP Retail, the determined requirement quantity (investment-buying quantity) is automatically taken into effect for the purchase order.

Stock allocation

Stock allocation comprises all functions for the distribution of a total procurement quantity to several recipients, normally stores. Thus, the stock allocation is primarily a means for the central procurement of articles for the individual stores. In addition, allocation tables are sometimes used for stock transfers or for the distribution of remaining stocks from distribution centers.

A stock allocation in SAP Retail can be defined using manually specified percentages or fixed individual quantities. Alternatively, distribution rules can be defined that are used to automatically distribute the total quantity to the different stores. The distribution rules and the differentiated control parameters (e.g., type of rounding for percentages, type of distribution for remaining quantities, distribution keys for insufficient quantities) often permit an automatic allocation. Thus, this tool can significantly reduce the manual (re)processing effort.

Order transmission and order monitoring

In addition to electronic purchase orders (e.g., via EDI), fax purchase orders dominate the order transmission in practice. Telephonic purchase orders are usual in the object business in the wholesale sector, because often parallel terms of delivery and conditions are agreed. As part of an external optimization of the supply chain, EDI concepts should be favored, because they are the only procedure for which the data can be passed without media fragmentation and without manual intervention from the retailing company's merchandise management system to the vendor's ERP system. The low level of penetration of EDI still prevalent in many retailing areas can be explained primary by the high installation effort resulting from the required bilateral agreements (e.g., as result of various EDI subsets and innumerable versions). In particular, retail companies that maintain delivery relationships to manufacturers from different industries must normally support several subsets.

The order monitoring has the task to check the observance of the agreed delivery dates by the vendor. As aid, automatically initiated notes produced shortly before the planned delivery date and reminders when the delivery date is exceeded can be used. Although SAP Retail also provides these functions, they are seldom used in the retail sector.

3.3.3 Goods Receipt

Goods receipt forms the interface between the procurement process and the warehousing functions. Thus, it represents the retailing company's starting point for its merchandise logistics. The merchandise logistics concern the planning, execution and monitoring of material movements, and the information flows that coordinate them. In addition to the goods receipt, the internal merchandise logistics in the retail company cover all activities associated with warehousing, inventory management and physical inventory, the picking and preparation of the goods issue, and the physical goods issue itself. The goods movements can take place between the vendor and the retail site, between two or more retail sites (e.g., between distribution center and store) or between the retail site and customers. Correspondingly, goods receipts take place both in distribution centers and in stores.

Prerequisite for an efficient handling of merchandise logistical functions with regard to an efficient consumer response (ECR) (see Section 7.1) is the avoidance of manual activities wherever possible through the use of an integrated retail information system. SAP Retail reduces the effort for the manual acquisition of the documents that accompany the goods movements by generally defining references to the previous documents and which only need to be manually changed for deviations. Figure 3-26 shows an example that illustrates the references between the documents of the procurement process.

The control of the delivered goods, the updating of the associated inventories and the physical receiving or forwarding to the requestor for internal consumable material are performed during the goods receipts. The return of unsold goods and returnable transport packaging to the vendors can also be combined with the goods receipt. When a warehouse management system is used, transportation orders for the physical stock placement can be produced automatically following the goods receipt.

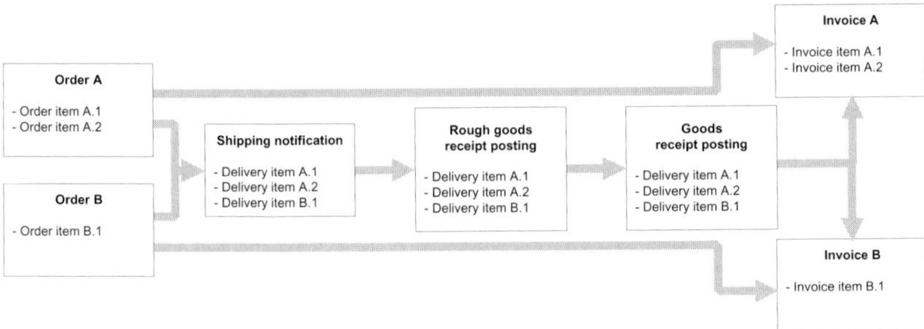

Figure 3-26: Document reference for the procurement process

Goods receipt planning

To increase the efficiency of the unloading operations, a goods receipt planning can be performed prior to the arrival of the expected delivery. In this case, a gate for the goods receiving at the associated warehouse complex is assigned to every expected delivery within the planned period. On the one hand, the knowledge of the docking gate simplifies a faster delivery by the supplier. On the other hand, it is the goal of the gate assignment planning to increase the ramp utilization and so to avoid bottlenecks. These conflicting goals, which, analog to the "scheduling dilemma" familiar from the production management, are called the "dilemma of the ramp assignment planning". Because no procedure exists to determine an optimum solution, heuristics must be used here. The system supports the planning side by providing automatically generated, but manually changeable, gate assignment suggestions and a graphical display of the gate assignment in the form of bar charts.

Goods acceptance/goods receipt recording

The processing effort for the acquisition of a goods receipt can be reduced through the reference to one or more existing purchase orders. If the corresponding order number is not known to the processing clerk, the associated purchase orders can be sought based on the vendor, the delivery date or the delivered articles. The individual delivery items suggested from the purchase order can be confirmed or changed at the goods receipt. An additional data acquisition is necessary only for articles delivered without a corresponding order item.

If a notification from the vendor with details of the delivery date and the delivered items precedes the delivery itself, this shipping notification with reference to the purchase orders can initially be entered and used as reference for the subsequent goods receipt. When stores are supplied from a distribution center, the goods receipt can also be performed with reference to the goods issue posting of the distribution center. In general, goods receipts can also be performed without preceding purchase orders. As option, the system can create a subsequent purchase order in this case.

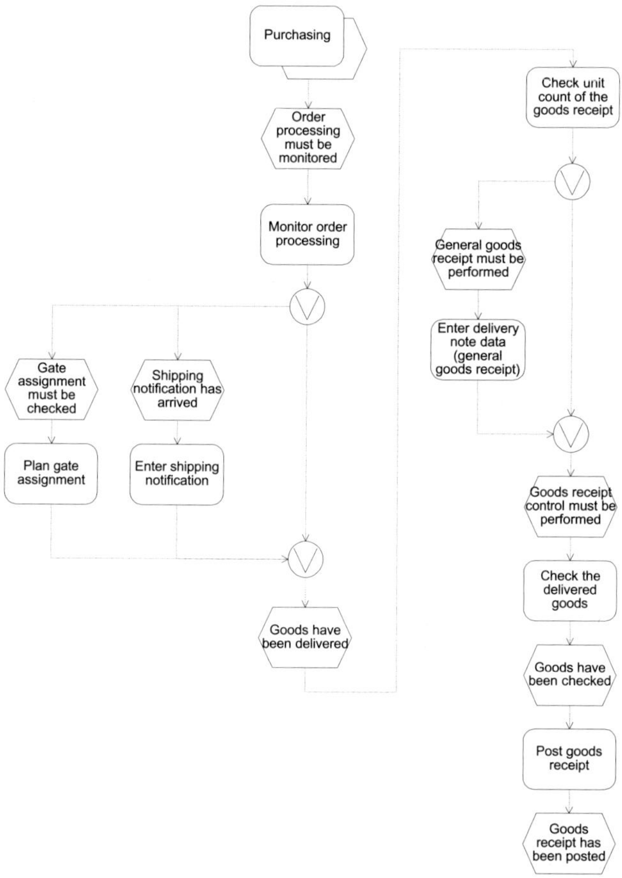

Figure 3-27: Goods receipt process model (simplified)

SAP Retail allows the goods receipt to be performed in one or two steps. In the two-step goods receipt, the "rough goods receipt" performed initially records the relevant header and delivery note items. In addition, based on the previous rough goods receipt the subsequent functions can be prepared earlier with the printing of count lists, sales labels and goods receipt/issue slips. The control and acquisition of the delivered goods with regard to type, quantity, possible exceeding of underdelivery or overdelivery tolerances and the observance of specific shipping and packaging rules is performed at the subsequent actual goods receipt. The two-step goods receipt and thus the possible time-related decoupling of the acquisition and production of documents, on the one hand, and the goods receipt control, receiving and booking, on the other hand, improves the planning of the goods receipt and thus avoids bottlenecks. Figure 3-27 illustrates the general procedure for the goods receipt processing.

Typically, other transaction types must be handled in retailing companies in addition to the receiving of the ordered items. The return of material to the vendor can also be performed as part of the goods receipt. This can be returns of unsold articles or discontinued articles, or returnable transport packaging. SAP Retail provides special transaction types for return deliveries. These transaction types are then used for an adjustment posting of the goods from the associated inventories.

Goods receipts in the stores can be posted directly in the central SAP system using the remote access or the SAP *Retail Store* system (see Sections 4.3 and 4.4). In addition, an interface can be used to transfer data from external store retail information systems or mobile data acquisition units. Then, the validation of the correct document reference and possibly the matching with the goods issue posting of the distribution center can be performed in the central system.

3.3.4 Invoice Verification

The invoice verification has the task to check the incoming invoices for the correctness of content, arithmetic and price. The invoice verification, in particular, underscores the advantages of an integrated retail information system: the previously mentioned checking tasks require information about the ordered goods (purchasing), the delivered/received goods (goods receipt) and the conditions agreed with the vendor generally and for the particular order. The data are all integrated in the SAP system. The conditions result from the purchasing info records and the pricing scheme, the ordered article quantities come from the purchase order, and the received delivery quantities come from the goods receipt postings.

In the ideal situation, the invoice data are transferred from the vendor by EDI. In this case, they can be added directly to the SAP system without needing any manual processing. If, however, they are provided in paper form, at least the invoice heading and the total amount must be entered manually. When no direct checking of the invoice takes place here, this is referred to as a *preliminary posting of invoices*. Manual or automatic procedures are used to determine the associated purchase orders and goods receipts. The automatic invoice verification

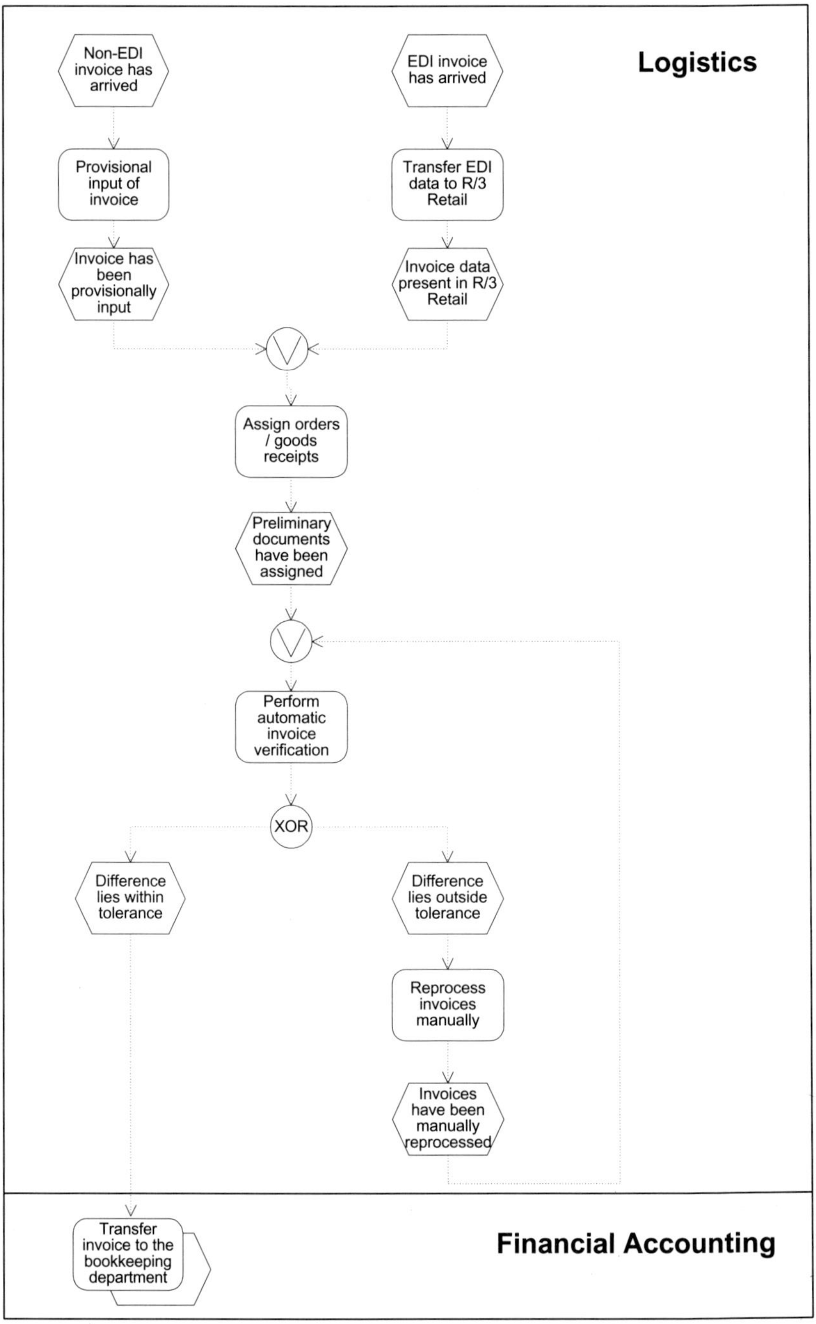

Figure 3-28: Logistic invoice verification in SAP Retail

then takes place. Provided the invoice amount determined by the system agrees with amount on the vendor's invoice (or lies within the tolerance limits), the invoice is accepted as being correct and is passed to the accounting department. If, however, there is a significant difference, this requires manual postprocessing.

The range of checking steps makes the automatic invoice verification a resource-intensive process. Consequently, a direct validation during the input can cause delays. In addition, many retailing companies separate the invoice entry from the postprocessing of incorrect invoices into different organizational units. Thus, in addition to the dialog-based invoice verification, SAP Retail permits a cyclically initiated verification as background job. Following the verification run, the invoices to be checked manually can be chosen using selection criteria. They can then be processed online.

Compared with the initial invoice verification process in the standard R/3 system, the invoice verification in SAP Retail is characterized by being separate from the financial accounting. Only after an invoice has been checked successfully is it passed for further processing to the financial accounting (FI module). SAP calls this innovative invoice verification, which is now also available in the standard R/3 system, *logistic invoice verification*. The decoupling from the financial accounting permits the logistic invoice verification to perform a decentral invoice entry and verification (e.g., in subsidiaries or distribution centers) without needing direct access capabilities to the central SAP FI system. Figure 3-28 shows the process for the logistic invoice verification in SAP Retail.

In accordance with the retail H-model, the settlement and monitoring of the purchase-side subsequent conditions (so-called bonuses) are assigned to the invoice verification. Because the same content-structural task also occurs on the sales-side (bonuses or commissions for customers), SAP Retail provides a central component, *subsequent settlement*, that can be used to manage and execute both the purchasing-side and sales-side subsequent settlements. This component includes the following functions:

- The creation and updating of the subsequent conditions agreed with vendors and customers. It supports conditions that are settled just once and also those settled periodically. Subsequent conditions can be assigned a validity period. The reference base (e.g., a quantity, a value, a weight or a number of points) can be selected freely.
- Support is also provided for a retroactive entering of subsequent conditions. Past turnover is determined and used to calculate the conditions.
- An update of the turnover relevant for the subsequent conditions can be made automatically with the purchase order, during the goods receipts or during the invoice verification.
- An intermediate settlement of the accumulated condition entitlements is possible at any time.

3.4 Warehousing

The warehousing allows the retail company to satisfy its time-related, quantitative and qualitative bridging function. A merchandise management system to support the warehousing functions is used primarily for the quantity and value representation of the warehouse stocks by entering stock changing actions and by the use of different valuation procedures. Furthermore, retail information systems can provide support for the preparation of the physical activities associated with the warehouse logistics, e.g., by creating inventory counting documents. Given the appropriate compatibility and connection, the retail information system can also initiate activities for the automatic warehouse control and transport systems, and thus further reduce the number of manual tasks in the goods logistics.

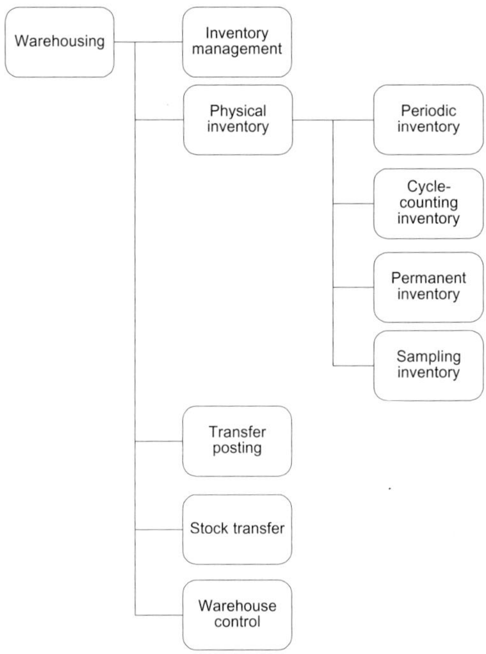

Figure 3-29: Warehousing function model

Figure 3-29 shows the functions (without goods receipt processing and goods issue processing) performed in the warehouse. The main functions are discussed in the following sections (see also Mulcahy (1993)).

3.4.1 Inventory Management

A representation of the inventory in the system is needed for planning and valuation purposes. Generally, the following means are available to determine the current stock level:

- record the physical inventory
- equate goods receipt and goods issue
- estimate the warehouse stock levels
- update the stock levels by recording the goods movements that change the stock level (balancing).

Integrated information systems have the advantage that given a correct acquisition of all inventory-relevant activities, not only an automatic balancing is performed, and thus always complete and current information about the level of all (target) stocks are available, but they also draw immediate attention to the affects on the previous and subsequent areas of the company (e.g., with the corresponding postings to the accounts of the internal and external financial accounting). This also permits the comparison of the stock levels recorded during the physical inventory with the booked target stock. Because of different uses, valuation procedures and ownership, the goods must be assigned to different stocks that have to be selected at the goods receipt posting. Figure 3-30 shows an overview of the possible stock types in SAP Retail.

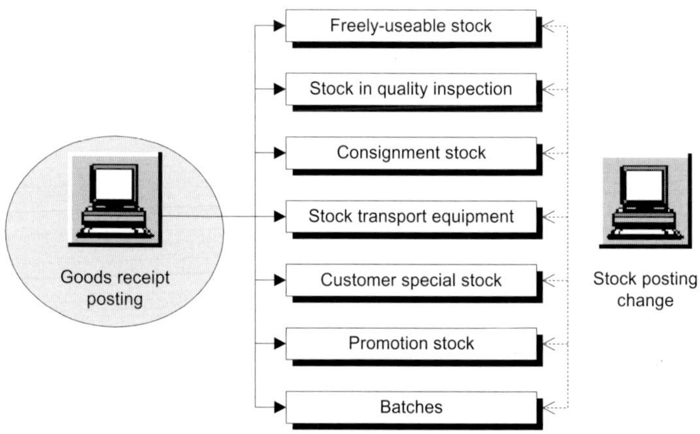

Figure 3-30: Overview of the stock types

The *freely usable stock* is available for the purchasing without restriction. Thus, perfect goods are normally booked to this stock. If goods after the goods receipt posting are not freely available immediately, for example, because no final quality check has been performed, a posting can be made to the *stock in quality inspection*. Similarly, goods that were accepted only conditionally, can be added to a *blocked stock*. Because *consignment goods* and *transport equipment* remain the property of the vendor, they are managed in special stocks and are not valued. In addition, *special stocks* can also be managed for customers as property owned by other parties. Similarly, an own stock of *consumables* is available for internal use. A specific *promotion stock* can be assigned to an promotion.

Articles to be batched can be maintained in stocks separated into *batches*. Batches are quantities of an article, which, as result of the production method, are not qualitatively fully homogeneous and thus require a separate handling. The inventory management in the retail site can be handled in two ways, either the batches allocated externally by the manufacturer can be accepted or new batches formed using internal criteria.

The differentiated inventory management resulting from the individual analysis capabilities, also for the preceding and following areas, permits an up-to-date and comprehensive overview of the various stocks. For example, the stocks can be analyzed according to quantity and value, the goods movements belonging to a stock (with access to the base transaction documents) can be displayed, and the balances of the goods receipts/invoice entry accounts can be analyzed. The stock analyses can take place at various levels: on the one hand, storage location dependent, and, on the other hand, at higher aggregation levels, such as warehouse area, site or complete company.

Structured articles, such as sets or displays, can be handled in two ways. Either the structured articles can be separated automatically into components when stock changes occur or individual articles can be combined to form structured articles. In many cases, an inventory management can be sensible both at the level of the individual components and also simultaneously at a higher level, such as when both types are sold in a store. The stocks of the individual variants of a generic article can be considered and analyzed together as a variant matrix.

The combination of empties and full products represents a special form of a structured article – with wide-ranging effects on the logistical processing (for example, a full beverages crate that consists of ten empty PET deposit bottles and ten bottle contents). The empties are initially created in the system as "normal" articles. However, they have the characteristic that a specific quantity of a full product article is always assigned to them which can be neither purchased nor sold without the empties. The connection of empties and full products is specified in a similar way as part structures in industrial operations, namely as bills of material, which, however, can currently have just a single level in SAP Retail. An automatic bill of materials resolution then allows access to the components of a full product during the processing of stocks.

However, for certain articles, an inventory management based on exact articles is neither useful nor possible. For example, an acquisition for exact articles during sales at the point of sale is not possible for technical reasons, is uneconomic or not of interest to the financial accounting and controlling. In these cases, the article stocks are managed purely in value terms. Typical examples for such articles considered normally only in value terms are fruit, meat or also hard goods when these are bulk materials, such as screws, etc. These articles are combined to form groups, which are designated as *value-only articles* and maintained as such in stock terms. The article groupings can be merchandise categories or other adequate groupings. The first case the value-only article is designated as merchandise category value-only article (see Section 3.2.4).

3.4.2 Stock Valuation

The value-related inventory management requires a valuation of the individual delivery note items as part of the goods receipt. Goods can always be valued using both their purchase price and their selling price. Retailing companies normally use both prices for inventory management. A goods updating using purchase and selling prices supports the short-term earnings analysis by determining the article-related gross margins and allows a judgement of the economics of assortments and the success of the managing organizational units (refer to page 123 for a discussion of the calculation of the gross margin).

In this procedure, the stocks are valued during the goods receipt acquisition. Thus, the goods updating is also designated as the determination of the goods receipt-related margin, because the purchase and sales values originate during the valuation goods receipts and only the sales department makes corrections resulting from special events. Examples for events that can cause a deviation of the realized selling price from the planned selling price and so require a correction to the calculated gross margin are price reductions for the sales of faulty goods and promotion sales.

In addition to the manual selling price revaluation, SAP Retail also permits an automated sales revaluation using selection criteria that identify all articles with changed sales conditions. If the selling price revaluation is performed margin-neutral, i.e., the gross margin should remain constant despite the correction, the cost price is corrected automatically.

Various approved procedures exist for the tax-related valuation of the stock that generally consists of articles with different purchase prices.

- The *FiFo* principle (first in, first out) assumes that those goods purchased first are also those sold first. Consequently, the valuation of the remaining stocks uses the prices of the most recently purchased goods.
- In contrast, the *LiFo* principle (last in, first out) assumes that the most recently purchased goods are consumed first. In this case, the stocks are valued using the prices of the goods purchased first.

- The *lowest value principle*, which corresponds to the principle of business conservative valuation, uses the lowest price pertaining during the balancing period for the valuation.

All three valuation procedures can be used with SAP Retail. Figure 3-31 illustrates the valuation of a warehouse issuance using the FiFo, LiFo and lowest-value principle, and the procedure of the moving average price.

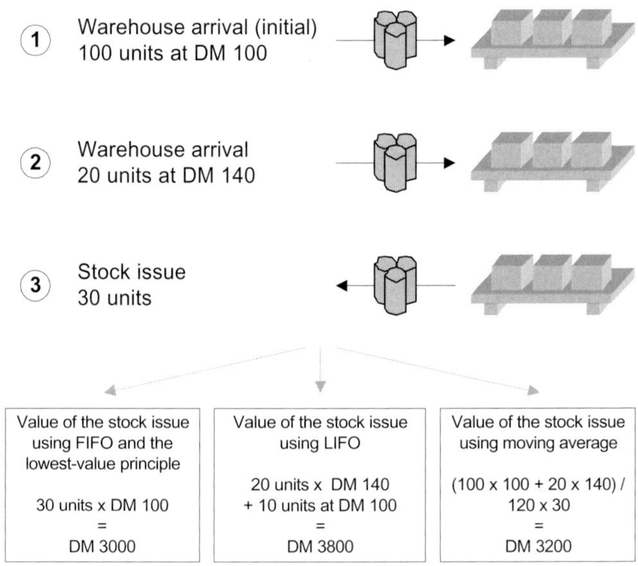

Figure 3-31: Valuation procedure in SAP Retail

3.4.3 Physical Inventory

The physical inventory involves the physical recording of the stocks at a specific time. The physical inventory is also necessary for a continuous updating of stocks to detect any losses not previously recorded. Many countries have laws that require the physical inventory to be performed at least annually. Differences between the target stocks and the physical stocks are designated as *inventory differences*. The major causes for inventory differences are errors during the acquisition of goods movements, goods losses resulting from theft, shrinkage or deterioration, and errors with the recording or documentation of the physical inventory.

A differentiation is made between the following physical inventory procedures:

- For the classic procedure, the *periodic annual inventory*, the recording of the stocks by measurement, weighing or counting is performed on the balancing day. Care must be taken here to assure both the completeness of the recording and the avoidance of double counting. Because it is often not possible to perform an exact complete physical inventory on the balancing day, the recording of the stock can also take place earlier or later and the value extrapolated to the balancing day. For example, the calculation of the goods stock value on the basis of selling prices for an earlier physical inventory is performed as follows:

	(stock at selling prices on the physical inventory day
+	goods receipt of the store at selling prices
–	store turnover during the period from the physical inventory day to the balancing day
–	usual shrinkage at selling prices during the period from the physical inventory day to the balancing day
–	usual physical inventory deficit between the physical inventory day and the balancing day)
*	(1 – average gross margin)
=	goods stock on the balancing day at cost prices

- In contrast to the periodic inventory, the stock recording for the *permanent physical inventory* can be spread over the business year. In this case, it is only necessary to ensure that each article is recorded physically at least once during the course of the year. Because both the periodic inventory and the permanent physical inventory apply to the complete goods stock, both procedures are also designated as complete stock counting methods.

- Under certain circumstances, the legal regulations are satisfied when only a part of the goods stocks are counted and used to extrapolate to the complete goods value. The fact that this method, known as *inventory sampling*, is scarcely used, although it, for example, has been approved in Germany for more than 20 years (§ 241 HGB), lies on the more rigid legal requirements. One the one hand, the chosen mathematical-statistical methods must satisfy the requirements of the generally accepted accounting principles, and, on the other hand, the meaningfulness of an inventory sampling must correspond to that of a complete stock count. The latter produces such large samples for subassortments heterogeneous in stock and value that the organization effort often exceeds the benefits of the lower counting effort – in particular in the retail sector (cf. AWV 1996).

SAP Retail also supports the *cycle counting* method. Here, the articles are recorded periodically in predefined time intervals during the course of the business year. These time intervals can be defined for specific articles and thus certain articles are inventoried more frequently than others. To support the short-term earnings analysis of the stores, an *intermediate inventory* can also be performed at any time.

Irrespective of the inventory method, the stocks for the article to be inventoried must be blocked for goods movements for the duration of the physical inventory. In the retail sector, this is achieved by performing a physical inventory outside business hours (or by closing the business to conduct the inventory). In contrast, in the wholesale sector, the warehouse areas to be inventoried are temporarily closed for stock placement, stock transfers and order picking tasks.

The acquisition of the physical inventory data can be made manually on inventory recording lists or with mobile data acquisition devices (MDA devices). SAP Retail supports both types of recording by creating inventory recording lists and counting lists, and allows the automatic transfer of MDA data using an interface. The manual acquisition of the physical inventory data in the system makes use of the previously created inventory documents. The system can list and adjust any detected inventory differences. Should the inventory differences be too large, recounting documents can be prepared for a new physical check.

Physical inventories are performed similarly in those stores that do not have their own merchandise management system and whose stocks are maintained in the head office system. The head office blocks the stocks to be inventoried in the system and inventory documents are directly printed or are sent with electronic data transfer to the store. The results of the store inventory are also transferred electronically or in list form to the head office where they are entered into the system and further processed.

3.4.4 Warehouse Control

In addition to the support of the inventory management and the physical inventory, the warehouse management tasks principally concern the preparation and execution of the physical stock placement and stock removal, and stock transfers. Warehouse control systems should guarantee the on-time, frictionless and economical processing of these physical goods movements. SAP Retail uses here the functionality of the SAP Warehouse Management System that is activated as result of the generation of transport orders and provides interfaces to automatic conveyor and fork lift systems, barcode readers, etc.

Various arrangement forms exist for the technical realization of warehouses: the warehousing in *block warehouses* is performed on the floor with weight-insensitive goods also being stacked. Because a direct removal in block warehouses is possible only at the edges of a block, and the possibilities of automation and optimization of the warehouse management are limited, *compartmentalized rack warehouses* or *pallet rack warehouses* are more often used in retail companies with a larger product range. *Pallet rack warehouses* are more suitable for warehousing a larger number of articles and can be more easily automated than the other warehouse types. High rack warehouses have not only the most complex structure of all rack warehouses, they also require the largest investment, management and control effort.

In addition to a goods-oriented structuring, larger warehouses are typically divided into various areas formed for functional aspects (e.g., goods receipt area, picking area, reserve warehouse, shipping area). *Reserve storage areas* provide space for articles that must be stored short-term in such large quantities that the reserved normal storage bin does not suffice for the article. This, for example, is the case for seasonal goods or for investment purchases.

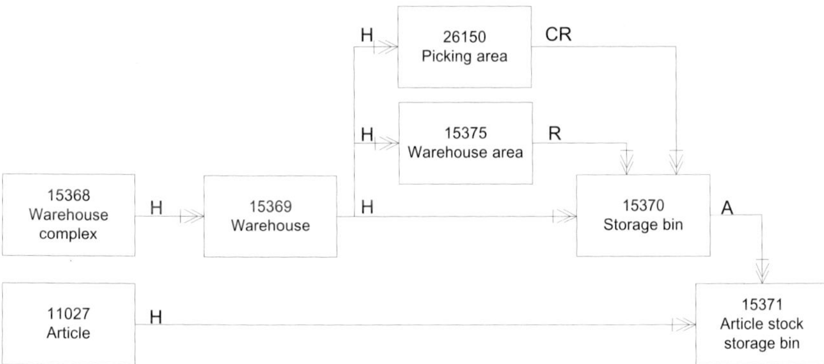

Figure 3-32: Warehouse organization in SAP Retail (simplified)

The technical form of the warehouse is initially not significant for the representation in the merchandise management system; what is required is an adequate representation of logical warehouse structure. Some of the constructs SAP Retail provides to describe the warehouse structure are warehouse complex, warehouse, warehouse area and storage bin (also refer to Section 3.1.1). The smallest special unit within a warehouse area is the *storage bin* that is referenced in the system using its warehouse coordinates. Further warehouse structures, such as shelves and shelves' level, can be emphasized by using an appropriate numbering of the storage bins.

Article stocks are normally maintained at the storage bin level. SAP Retail allows stocks of an article to be distributed over several storage bins. Similarly, the stocks for different articles can be maintained at one storage bin (e.g., small accessories in the picking warehouse).

Figure 3-32 shows the interaction of the warehouse structures using the SAP data model.

The coupling with the warehouse management system in SAP Retail allows automatic transport orders to be created for the physical stock placement when the acquisition of the goods receipt completes. Initially, the warehouse area and the storage bin are determined for the storage location of the received articles. In addition to the manual assignment of a storage bin, the SAP warehouse management system provides a choice of various stock placement strategies:

- use of *fixed storage locations* to which articles have a fixed assignment;
- stock placement in the vicinity of the fixed assignment at a *reserve storage bin*;
- *addition to existing stock* at storage bins that already contain the type of article to be stored (provided sufficient remaining capacity is available);
- stock placement in a *general storage area* in which only warehouse areas are differentiated but not further divided into storage bins;
- selection of the *next free storage bin* that the system suggests for chaotic warehousing;
- selection using the *warehouse unit type* (e.g., pallets) for which suitable warehouse positions must be located;
- *block storage* for articles with normally high number of items per delivery and correspondingly large storage requirement;
- use of a *customer-defined stock placement strategy*.

The use of a warehouse control system permits the automatic physical stock placement and removal. Otherwise the warehouse worker must perform this task.

3.4.5 Stock Transfer

The transport of goods from one storage location to another is called stock transfer. Thus, stock transfers contain a goods issue at the sending storage location and a goods receipt at the receiving storage location. The storage locations can belong to the same site or can be located at different sites (e.g., for stock transfers between distribution centers and stores).

For stock transfers between sites that belong to different company codes, the stock transfer is considered as being a sale and causes the corresponding postings in the financial accounting. In these cases, a selling price must be selected that can differ from the stock value. If this settlement price lies above the purchase price of the supplying site, these are earning activities in the two participating organizational levels. The specification of an appropriate settlement price is gaining in importance with the increasing autonomy of the stores and the decentralization of the decision-making competence. Such a settlement price represents a classic problem for the coordination of organizational units.

The SAP system allows either one-step or two-step stock transfers to be performed. For the two-step stock transfer, the receipt and issue of the restocked goods is recorded in two separate posting steps. Thus, the information about the two actions is retained. In contrast, only one stock transfer posting takes place for a one-step stock transfer. In both cases, neither purchase orders nor deliveries are produced in the system. In addition, as part of a normal order and shipping handling, including invoicing and invoice verification, a stock transfer can take place between two sites with different company codes, even without stock transfer posting.

Because a transfer posting involves just an assignment of articles to a different stock type (for example, by releasing goods previously held as stock in the quality inspection), pure transfer postings differ from stock transfers. Transfer postings are normally not associated with goods movements between different storage locations.

3.5 Distribution

Distribution covers all decisions and actions associated with the path of goods to the end customers. With marketing, sales, goods issue, billing and debtor accounting, the distribution process of the retailing H-model covers five central functional areas.

The *marketing* area covers, in particular, product policy (e.g., assortment decisions) and contract policy problems (e.g., price and discount policies). The central objects to perform these tasks are assortment and subassortment that result from the listing. All customer contacts starting with the inquiry, through the quotations and orders, finishing with the customer complaints processing are handled by *sales*. The *goods issue* covers route planning, order picking planning, order picking and preparation for shipping, etc. The goods are sent to the recipient (customers or stores) once the shipping preparations have been completed. Depending on the order type (cash sale or invoiced sale), the invoicing process takes place together with the goods issue or after a delay. When the invoicing process takes place later, the recipient invoices are produced and then the *debtor accounting* monitors and posts the payment arrivals.

As with the creditor accounting, the debtor accounting is not further discussed in the following sections. Not only are the requirements very homogeneous, but various financial accounting systems, including the SAP FI module, also cover them comprehensively.

3.5.1 Marketing

In addition to the updating of the customer master data, as already described in Section 3.2.3, marketing, which the retailing H-model considers primarily to be operative marketing, is responsible for the quantity and value planning of the articles available on the market (goods planning), the building of assortments (assortment planning), and the article listing. The selling prices are also determined as part of the condition policy. In addition to make (standard) selling prices for the individual articles, a differentiated recipient or recipient group related price or condition fixing is usual, in particular in the wholesale sector.

Goods planning

It must be determined during the goods planning which merchandise categories are to be maintained in which depth, which quantity sales are forecast for the article and which turnover is planned for the future periods. This basis can be used to perform the sales and turnover planning that provides the necessary data for the limit calculation (refer to Section 3.3.2). The goods planning can be performed at the level of the individual articles or at an aggregated level (e.g., merchandise category). The data from the goods planning allow an improved control and monitoring of the company. This is especially useful for the better anticipation of expected seasonal sales fluctuations.

DB scheme	Plan 2001	Unit	Plan 2000	Actual 2000
Sales quantity	20,000	Pieces	1,500	1,800
Revenue	100,000	$	75,000	90,000
Net revenue	80,000	$	55,000	65,000
Cost value	40,000	$	35,000	40,000
Profit margin	30,000	$	10,000	15,000

Figure 3-33: Planning scheme in the controlling

Since release 4.6A, SAP Retail provides a completely new developed goods and assortment planning that mid-term should replace the previous planning functionality. This goods and assortment planning is primarily based on a limit planning and the classic controlling functions (CO-PA module). Although, initially the old planning functions are retained, they will not be further developed.

In the conventional planning (CO-PA module), planned values (value and quantity data) can be set for articles, for article groupings and for merchandise categories.

This information and the actual values can be obtained from the integrated data basis of the SAP system for analysis purposes. Figure 3-33 shows a simple example of a planning data structure.

Various article classifications can be used for planning and analysis purposes. In addition to the article grouping and classification capabilities already described in Sections 3.2.1 and 3.2.5, SAP Retail provides the following criteria for article grouping:

- **ABC indicator**: Classification of the article using an ABC analysis. The "A" value indicates an important article, "B" an article of medium importance and "C" an unimportant article. A possible classification can be based on the gross margin, the sales figures or the strategic positioning of the article.
- **External merchandise category**: In addition to the merchandise categories of the retailing company, further merchandise category classifications exist whose representation can sometimes be very useful. For example, CCG or Nielsen have defined their own merchandise category catalogs. The provision of such external merchandise categories allows the retailing company to use Nielsen merchandise categories to determine its own turnover for analyses, for example, even though the company's own merchandise category structure does not correspond to the Nielsen structure. This is particularly useful for inter-company analyses (benchmarking).
- **Price article**: Grouping of articles having the same price. The conditions only need to be updated for the specified price article; the other articles reference the prices and conditions stored for the price article.
- **Price band category**: Individual classification into different price classes. The company's own entries are defined in the customizing.
- **Product hierarchy**: The product hierarchy is used primarily for analysis purposes and for pricing functions. The key is formed in the standard setting for SAP Retail from three levels. Each level has a fixed number of characters.
- **Division**: This characteristic can be used to further subdivide the sales area.

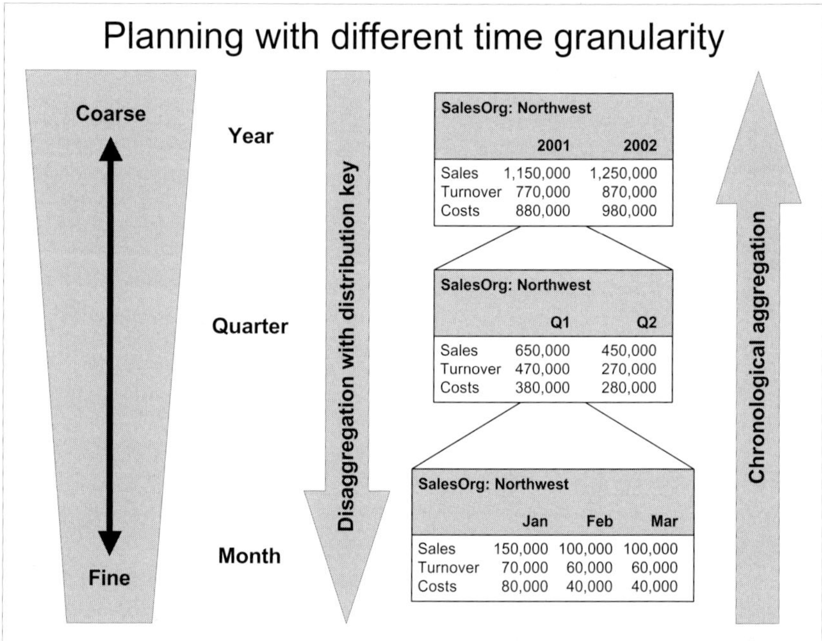

Figure 3-34: Distribution of planning data according to periods

The new SAP goods and assortment planning component is characterized by the following features:

- Definition of a *planning scenario* to structure the planning process
- Management of various *planned versions*; e.g., to compare alternative plans or to compare planning data with historical data
- Support for *value* and *quantity* planning
- Planning at an *arbitrary hierarchical level* (e.g., article, merchandise category, store group, purchasing organization)
- Saving of the planning data at the finest level so that consistent data are always available, even for *aggregations* and *disaggregations*
- Use of *time-based distribution keys* for the automatic distribution of planning data to individual periods (refer to Figure 3-34)
- Direct *EXCEL connection* using OLE (Object Linking and Embedding)

Figure 3-35 shows a typical planning process. Initially, the goals are defined (1). These include, in particular, the strategic goals formulated by the management (e.g., increase the turnover annually by 10 percent or extend the store network by making 20 new openings every year). In the goal planning (2), the strategies are made concrete with regard to the value for coarse criteria (e.g., purchasing organization, distribution chain). Head office key figures that need to be planned include, for example, sales, shrinkage and markdowns.

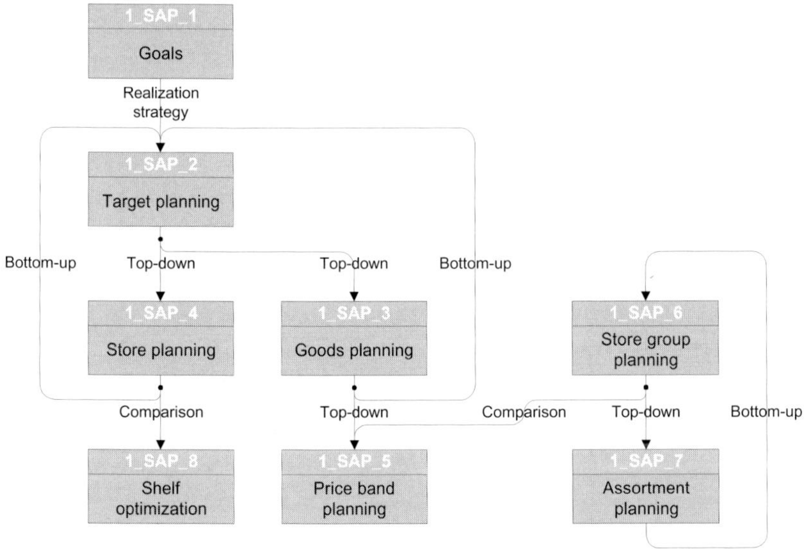

Figure 3-35: Procedure for a goods/assortment planning in SAP Retail

The data defined in the target planning are broken down into store level or merchandise category level so that they can be used as basis for the store and goods planning (3) (4). To ensure a matching can occur between the planning levels, the store and goods planning should use the key figures from the goal planning. The gross margin and the stocks always have to be planned. Although a value-related stock planning is performed during the store planning, it can also be performed quantity-related as part of the goods planning. Further key figures for the stock planning can be, for example, initial stock, purchases, sales or returns, from which further key figures, such as the range of cover or the *inventory turnover* can be calculated automatically. The next refinement level after the store planning is the shelf planning/optimization. For this purpose, a direct branch can be made from the planning scenario into an external shelf optimization program (8) (e.g., *SpaceMan* or *Apollo Professional*; see Section 7.2.2; Schütte/Vering/Wiese 2000, page 290 and following pages provide an overview of the leading shelf optimization systems and their providers). The goods planning (5) refines the price level planning, which, in particular in the fashion area, belongs to the standard planning methods.

In the store group planning, in which the stores can be grouped according to their turnover or other criteria, the assortment width and depth, and thus the number of articles, are defined (6). To ensure that the planned assortment form suits the planned quotas of the individual price levels, a matching can (and should) take place between the store group planning and the price level planning. The result of the store group planning enters the assortment planning (7). Therefore, the planning data are broken down to the article level.

Assortment planning

The major functions of the assortment planning and management are:

- the specification of the articles that are to be sold in a store
- the specification of the distribution centers used to supply the stores (by assignment of the article assortments to distribution centers)
- the specification of the stores in which the article is to be sold
- the determination of the customers for which the article should appear in the assortment (wholesale sector)

To ensure that the range of articles can be handled, the *assortment modules* used in SAP Retail represent an article grouping for assortment management and listing purposes. However, an article is not assigned directly to an assortment module, but rather assigned to an assortment module item. This construction allows the time to be limited for which the assignment remains valid.

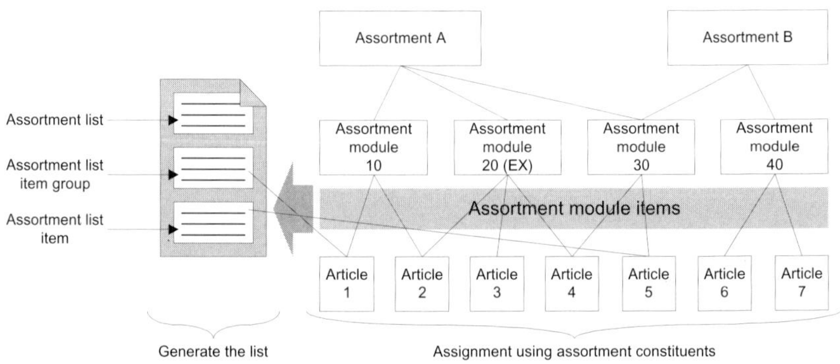

Figure 3-36: Overview of the assortment structure

In addition to this time-controlled assignment of an article to an assortment module, the same article can also be present in different assortment modules at the same time (cf. article 4 in Figure 3-36). In the second step, the assortment modules are then assigned to the individual assortments. An assortment module can be contained in several assortments.

However, the described type of building the assortment using a simple addition of assortment modules (so-called normal modules) is not the only possibility. If certain articles are to be removed from an assortment, this can be done using an exclusion module. Because an exclusion module has the highest priority, all articles contained in such a module are definitively removed from an assortment when the exclusion module is assigned. In the assortment A, shown in the previous figure, article 2 initially enters with module number 10, however, this assignment will be later discarded through the exclusion module (number 20).

SAP Retail provides various module types for different purposes. The most important are:

- **Normal module:** Freely selectable combination of articles for assignments to assortments. Such an assortment may be assigned to all stores. Normally, long-life articles or articles always available for all stores (e.g., important branded articles) are added to these modules.
- **Local module:** This is a special case of the normal module that can be assigned to only one assortment.
- **Exclusion module:** An article available in an exclusion module will be removed from the assortment, even when it is contained in other modules of the assortment. The exclusion rule overwrites all other assignments and cannot be disabled by other rules.
- **Promotion module:** These modules are formed for promotion articles. In all other respects, promotion modules are structured analog to normal modules.
- **Rack-Jobber module:** These modules are assigned directly to a vendor. Prerequisite for the combination of article and module is that the vendor delivers the article in the rack-jobber procedure to the store. A rack-jobber (shelf wholesaler) is a wholesaler or manufacturer that is provided with shelf areas (or a larger part of the sales area) in the stores. SAP Retail handles rack-jobbers as vendors. The ownership of the goods remains with the rack-jobber until they have been sold. A rack-jobber article can only be assigned to just one such module.
- **Value contract module:** This module owns an assignment to a customer value contract in which acceptance quantities can be specified.
- **Profile module:** These modules are created automatically from the article updating and thus cannot be updated directly. They are required as classification of merchandise categories for the assortment creation and management.

Assortment lists can be generated on the basis of the assortment, which has been defined using the assortment modules (refer to Figure 3-36). All articles of an assortment are maintained in a list. The generation can be performed automatically in regular intervals or initiated manually.

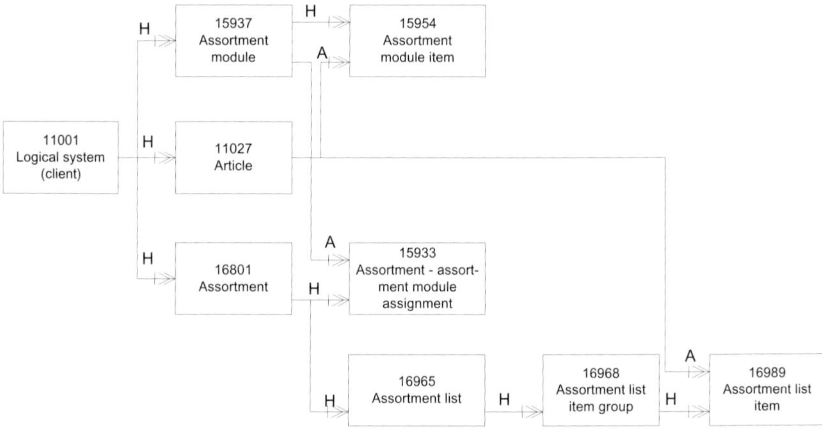

Figure 3-37: Data model assortment (extract)

Figure 3-37 represents the basic structure of the assortment in SAP Retail as data model. Furthermore: an assortment is assigned to just one distribution chain. However, a distribution chain can cover several assortments.

Article listing

The actual article listing, namely the assignment of articles to recipients, can be performed only after the sales views of the articles (in the goods planning) and the structure of the assortments (in the assortment planning) have been updated.

The listing is performed in SAP Retail using a direct assignment of recipients (stores or customers) or recipient groups to the previously defined assortments (refer to Figure 3-38). Just one assortment is assigned to a retail site (store or distribution center). In contrast, several assortments can be assigned to customers. The updating and assignment is performed manually.

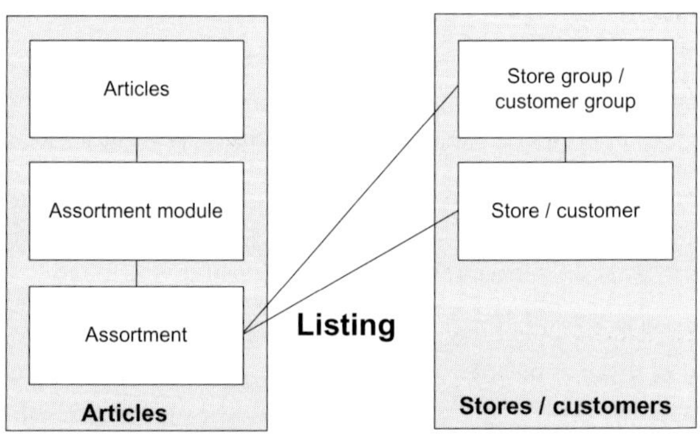

Figure 3-38: Listing in SAP Retail

Condition policies

The prices for the recipient are defined within the condition policies. Whereas the prices are normally identical for all recipients in the retail sector, the wholesale sector is characterized by having differentiated recipient-related prices and conditions.

The business processes of the wholesale sector normally require the inclusion of scale prices at the sales side. On the one hand, these should provide an incentive to place purchase order with larger quantities (with the goal of absolute profit increase despite a relatively low profit margin per item). On the other hand, the transaction costs associated with the handling of the sales order are lowered for the retail company through the reduction of the number of orders. This allows part of the cost advantage resulting from scale prices to be passed on to the customers.

Analog to the purchasing side, the price and conditions definition at the sales side can be made at various levels. A definition of the base article prices is normally made at the level of the distribution chain. These prices and conditions then can be overwritten by more specific (e.g., customer-group-related or customer specific) prices and conditions.

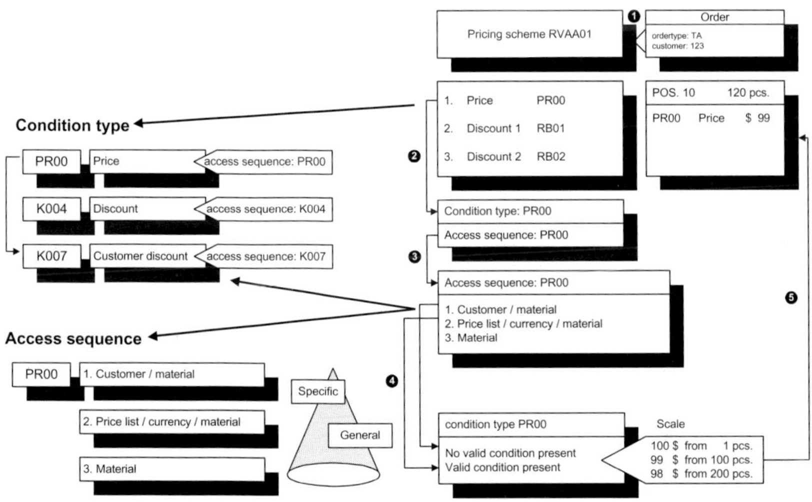

Figure 3-39: Determine conditions in SAP Retail

The assignment to a special price article can simplify the updating of the article prices (refer to Section 3.5.1). Thus, for example, a generic article "chocolate, medium price segment" can be defined with a standard selling price. This price article is then used to determine the selling prices of all relevant chocolate variants, even from different manufacturers. If the prices for all chocolate variants change, it suffices to change just the reference price.

The updated article prices are used as starting point for sales price calculations. However, the various markups and markdowns warehouses (cf. Diamond/Pintel 1996, page 234) or scaled discounts can cause the prices in the concrete business transaction to change. Consequently, the starting price after taking account of the available scale prices and supplemented with specific customer, article or sales line-specific markups or markdowns is used as input to the calculation of every item within a quotation, an order or an invoice. Freight and tax markups can also be taken into account.

The example in Figure 3-39 shows the processing to determine the conditions in SAP Retail. The starting point is the *calculation scheme* that results from the combination of customer and sales document type (1). A calculation scheme contains different *condition types* (and their order of calculation). A test is made for each of these condition types whether an *access sequence* is present in the system (2). The access sequence specifies in which order a search for a valid value must be made (3). The first valid value that is found terminates the search. In this example no price is found for the first entry of the access sequence (customer/ material). This means no customer-specific price exists. A valid condition is found based on the second entry of the access sequence (price list / currency / material) (4). Because this is a staged condition, the valid staged level based on the order item is determined (5). This produces a price of $99.00 in this example. However, because the same procedure must still be performed for the discounts maintained in the calculation scheme, this is not yet the end price.

The described sequence makes it apparent why a base price should always be defined for each article: if the SAP system cannot find a more specific price, the base price is used. If no base price is available, price details must be specified manually.

3.5.2 Sales

The sales department processes queries and orders as incoming customer contacts. The creation of order records, which includes the creation and the transfer of recipient order documents, is a further subfunction.

Customer query processing

SAP Retail differentiates between two types of customer queries:

- **Special queries:** These always apply to concrete articles already present in SAP Retail. These are typically higher priced products for which specific price details are required or articles with highly fluctuating ruling prices (e.g., electronic components or raw materials). Quotations can be produced in the further processing for these type of queries. System support is required for these customer contacts, because this is the only way to ensure that the prepared quotation and a possible order refer to the same products and the same prices. The automatic transfer of the individual items from the query into the follow-on documents avoids the need to reenter the data.

- **General queries:** General queries do not have any direct reference to the retailing company's articles. For example, a general request can be a request for an overview of a product program (a catalog) or a general query about properties of certain products or product groups (e.g., size, weight, dangerous goods class).

Create order record

Once the articles have be prepared in the merchandise management system for the sales department and valid reference paths for customer-article combinations have also been created, the create order record function can be used to produce order documents for the individual recipients (customers or stores). The *order record* shows the recipients which articles (in which logistical units) they can order.

The order record can be a paper-based list or an electronic document that is transferred by EDI. A digital order record can also be transferred from the central SAP system directly to MDA devices. This, for example, is useful for retailing companies if the stores order manually. This allows the store orders to be made directly at the shelf using an MDA device (e.g., for planning by visual inspection). Once all order items have been entered, the data are also transferred digitally to the central SAP system. This permits the order entry to be made in the store (even in case of a planning by visual inspection) without requiring any paperwork.

SAP Retail represents order records as *assortment lists.* Such lists can be created separately for individual customers or customer groups. Because the issuing units can also be specified, the combination of customer and article order shows in which logistical unit the customer has ordered. As soon as the order records have been created, they can be sent by post in paper form or as CD-ROM, transferred by fax or transmitted using EDI. When digital transmission (EDI) is used, the order record data can serve as basis for the master data input of the store retail information system.

The classic order record is increasingly being augmented with electronic catalogs and has already been replaced in some areas (technology and electronic assortments). SAP Retail provides a special product catalog component that can be used to create classic product catalogs and store-internal brochures, CD-ROM catalogs, multi-media kiosk systems or virtual stores for the Internet (refer to the discussions on SAP Online Store in Section 4.5 for the latter). Such product catalogs, in addition to the data from the SAP Retail system (article master data, prices, assortments, etc.), can also contain data from external sources (e.g., text, organizational structures, pictures, sound) (see Figure 3-40).

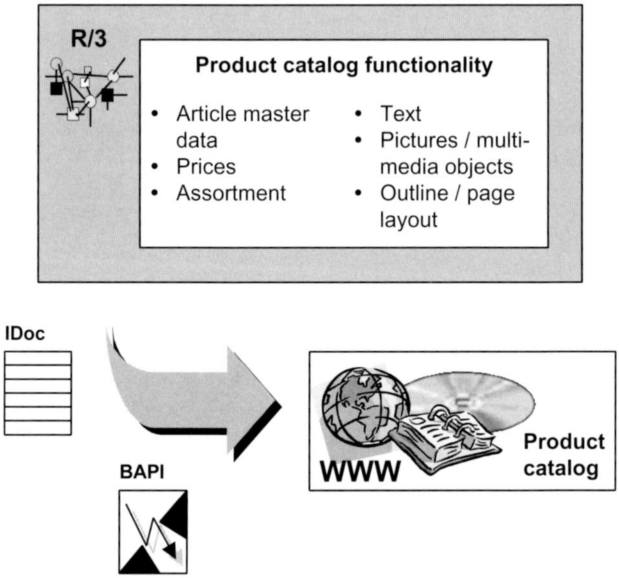

Figure 3-40: Creation of a product catalog using SAP Retail

Customer quotation processing

In SAP Retail, the entry of quotations is built similarly to the entry of the preliminary queries and the subsequent order entry. The crucial part of the quotation creation is the input of the article numbers (or the search for articles using various entry points) and determining the article price while taking account of the associated quotation quantity. In addition, the delivery date can be notified after making an availability test. The quotation data and the determined prices are stored. This ensures that the details from the quotation are available for a subsequent customer order. This aspect is particularly important for the quoted prices, because legal requirements require that the offered prices are observed for a certain period of time.

Because the functionality of the quotation processing largely corresponds to that of the order processing, the following section describes the provided functions in more detail.

Order processing

SAP Retail can be used to process both sales orders from the wholesale sector and sales in the retailing sector, although in different ways. Incoming sales orders from the retailing sector normally originate from the POS processing. The data from the

POS or the store retail information system (normally goods movements, money transactions and statistical data) are transferred subsequently to the (central) SAP system as past sales.

In contrast, sales orders in the wholesale sector are entered online at the SAP system. The sale has not yet taken place for such a sales order and the information needed to perform the business transaction must first be determined. Orders do not need to be entered directly at the head office; terminals that also allow stores to make order entries can be used here. The data entry input screens always have the same structure independent of the technical connection (directly over a LAN or with a terminal connection in the WAN).

The order entry normally starts with entering data into the document header, such as the name of the customer. A role resolution is performed in background using the customer number to determine whether there is a different goods and invoice recipient or payer. If other alternatives (e.g., different goods recipients) are found, the user must select the required alternative for the concrete business transaction.

The central inputs of the item data entry are the article number and the article quantity. Other details from the article master data (for example, article text or the standard quantity unit) are displayed automatically using the article number. Provided the data have been previously entered in the article master data, SAP Retail can also determine the article weights and sizes. Weights and volumes can be critical for transport planning and execution.

The pricing is performed online using the previously discussed calculation scheme (see page 87). Provided the processing clerk has been given the appropriate authorization, he can overwrite all suggested or determined data. Typical changes are the reduction of the starting price or changes to the discount rates.

A direct availability test can be performed during the order entry. If required articles are not available, it may be possible to use previously defined substitute articles. If it is determined during the test for availability that partial deliveries are needed for an order item, because, for example, only a limited quantity is available at the desired delivery date, the order item can be further subdivided using *classifications*. If customers do not want partial deliveries (or specify a maximum number of partial deliveries), this can also be specified during the order entry.

If the entered data for an order are changed afterwards, the system tests the consistency of the previously determined intermediate results and automatically performs any required recalculations and changes. For example, a modification of the international commercial terms can produce a new calculation scheme, such as when changed custom duties are to be used. Similarly, a subsequent change of the goods recipient causes the associated delivery site or shipping point to be determined again.

Such examples provide a good illustration of the dependencies involved in the order processing. The integration of all areas of a company in a system also per mits a direct branch from the order processing into the customer master data or the transaction data for a forwarding agent or to display the debtor data for a customer (e.g., open items).

Once the entered order has been stored, the subsequent functions, delivery and invoicing, are available immediately. To allow the various sales proces (type of the test for availability, time of the delivery, type and time of the invoicing, different form/document control, etc.) to be represented adequately, SAP Retail differentiates between various order types. The three central order types are:

- *Standard order*
  The order processing using a standard order assumes a time separation between the order processing, the delivery and the invoicing.
- *Rush order*
  For a rush order, the customer fetches the goods immediately or the retail company delivers to the customer on the same day on which the order was issued. As with the standard order, the customer pays later. When the rush order is stored, the system suggests the current date as delivery date. A delivery is created in the background when the order is posted.
- *Cash sale*
  For a cash sale, the customer himself fetches the goods, and, in contrast to the rush order, pays immediately when the order is given. Consequently, the order quantity, the delivery quantity and the invoiced quantity match. When a cash sale is stored, the system suggests the current date as date for the delivery and invoicing. When the order is posted, a delivery is produced in the background, the invoicing run is started and a cash sales invoice is printed directly.

SAP Retail also provides various additional order types for special handling (e.g., fetch returnable transport packaging, free delivery). Additional special order types can be defined during the customizing.

3.5.3 Goods Issue

The goods issue represents the removal of goods from the site stock for the supply of a recipient, for internal consumption or for the destruction of non-saleable goods. The recipient can be either a customer (thus, the initiation of the goods issue is a sales order) or another site of the retailing company, e.g., a store that is supplied by a distribution center. The central functions that a retail information system must support here include the planning of the routes to supply the recipients, the picking of the goods, the shipping handling with the packaging of the goods and the posting of the goods issue. The retail information system must monitor the dates for the shipping and has to initiate the picking, packaging and on-time shipping by using backwards scheduling. The handling of customer returns that follow the shipping of the goods is performed as part of the goods receipt processing. Figure 3-41 shows the general goods issue process.

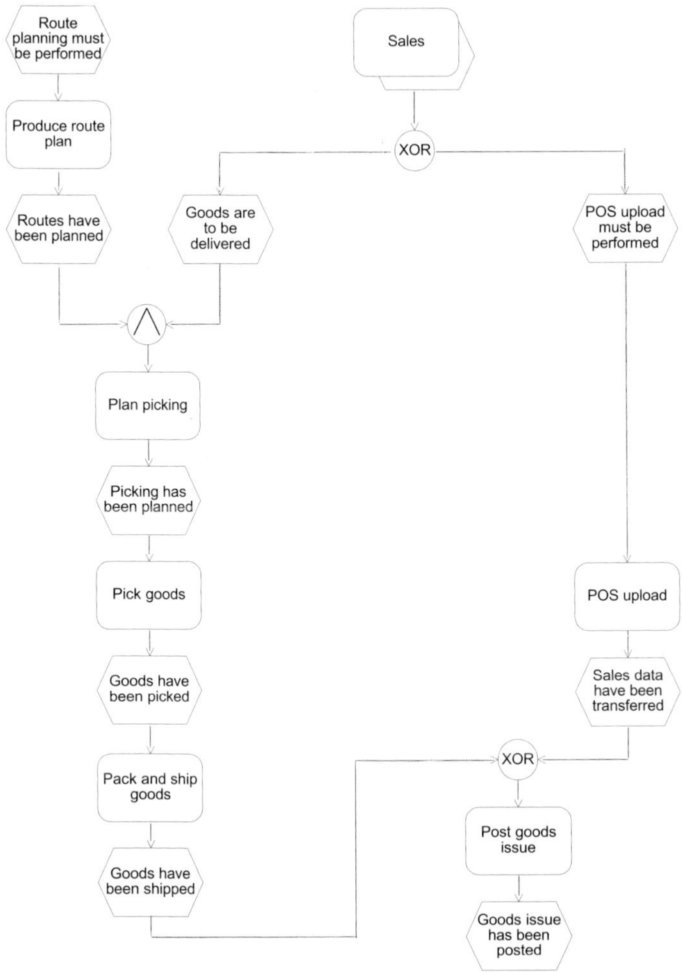

Figure 3-41: Process model goods issue DC to store (simplified)

Route planning

The route planning requires that the deliveries are assigned to the available means of transport so that they not only reach their recipients on time, but also as efficiently as possible, i.e., normally using routes that have the minimum length. A route plan always contains information about the means of transport, the geogra phic route, the departure date and the departure time. The determination of the route planning involves assigning suitable route schedules and specifying the itineraries to the planned deliveries. The planned delivery times are calculated using the present departure time and the expected journey time to each of the goods recipients. If a recipient can accept the goods only at certain times, this restriction must be used in determining the itinerary. The system can suggest the route planning and the itinerary, which, however, can be modified manually by the user. Such a procedure is called static route planning, because the sales orders, and thus the volumes to be delivered, are known for the journey planning. Because many recipients are in any case supplied in fixed cycles, such a planning is adequate in most cases.

Pick planning / picking

Picking designates the combining of the articles for the orders, which takes place in a warehouse. The removal of the goods is made either by a warehouse employee ("man-to-goods" picking), possibly using manually operated transport vehicles, or with an automatic transport system that transports the goods to a stationary picker and the remaining goods are returned to the warehouse after removal ("goods-to-man" picking). For in-store sales, the picking is not performed as a warehouse removal but by the customers themselves at the shelves in the store.

To optimize the use of resources for the picking and to avoid bottlenecks, picking orders can be combined to form *wave picks* that each can be assigned to a route planning journey. The scheduling of the picking must then ensure that the wave picks do not overlap and can be completed before the start of the associated journey. This requires that the picking quantities, the handling times per article, and the set up times are known.

A two-stage picking can achieve a better use of the available capacity when initially the complete required quantity for an article is picked and is distributed to the various sales orders in the second step. In comparison, in case of single-stage picking, the goods are removed directly for each individual recipient. The picking for extensive sales orders that require fast processing can be accelerated by creating several picking orders, which then can be processed in parallel in different warehouse areas.

The combination of the single and two-stage picking with such parallel or serial picking procedures produces the four possible organizational forms shown in Figure 3-42. These differ in the resource utilization, in the throughput times and in the coordination costs.

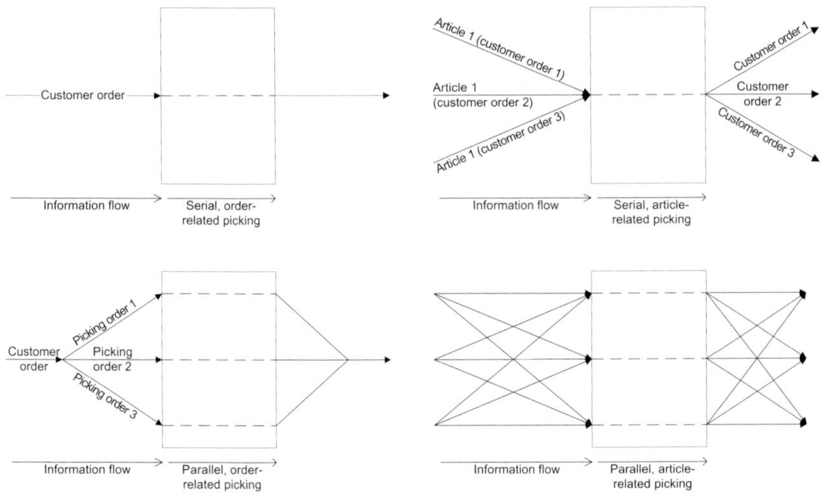

Figure 3-42: Picking organizational forms

If the stocks for an article to be picked are contained in several storage bins, different stock removal strategies can be used to search for a storage bin for the picking. In addition to the LiFo (last in, first out) and FiFo (first in, first out) methods, it is also possible to base the stock removal on the earliest use-by date or the quantity to be picked or to use a customer-defined stock removal strategy.

If a picking area has a fixed bin assignment in which articles are always stored in fixed storage bins, a replenishment planning can be used to automatically prepare the stock transfer from the reserve warehouse. Additional information about the planned deliveries for the affected fixed bin article is used for the replenishment planning.

Shipping

The articles must normally be *packaged* for their shipping. In this case, the individual deliveries and delivery items can be combined to produce *shipping elements*. Shipping elements can represent cartons, crates, pallets or trucks, and can contain not only the articles but also other shipping elements. The grouping of these shipping elements does not need to agree with the grouping of the individual deliveries: Whereas delivery items for a delivery can be packed in different shipping elements (e.g., in several cartons), deliveries for different recipients can be assigned to a single shipping element (e.g., truck).

SAP Retail can manage the shipping materials used to pack the items as separate articles with reduced master data.

Goods issue posting

Finally, the goods issue is posted in the system with the associated consequences for all affected areas. The stocks for the delivered goods are reduced, the required postings are made in the financial accounting and cost accounting modules, and the billing is initiated. The goods issue postings for sales in the stores are being performed afterwards with the import of the sales data (POS Upload). SAP Retail can process POS data from different POS systems. Suitable converter programs are available for most POS systems (see Section 4.7).

3.5.4 Billing

Billing covers all functions associated with the producing and processing of the invoice. In the retailing sector, these functions take place at the POS (refer to the reduced retailing H-model for the retailing sector; Figure 2-4, p. 15). Consequently, the following section assumes the typical procedure for the wholesale sector, mail-order and Internet retailing.

Recipient invoice creation

The billing completes the business transaction from the viewpoint of the sales department. When orders can be billed directly (e.g., cash sales), the invoice is created prior to the delivery. However, normally, the invoice is created with a time delay after the goods have been delivered. In this case three different invoice strategies can be used:

- A *standard invoice* bills one celivery (1:1-relationship between delivery and invoice.
- A *collective invoice* bills several deliveries with a single invoice. The deliveries may have been produced from different orders.
- The *split invoice* represents a third possible way to create invoices. A split invoice involves producing several invoices for a single delivery note.

Figure 3-43 summarizes these basic forms of producing the invoice for the recipient. SAP Retail supports all of these forms.

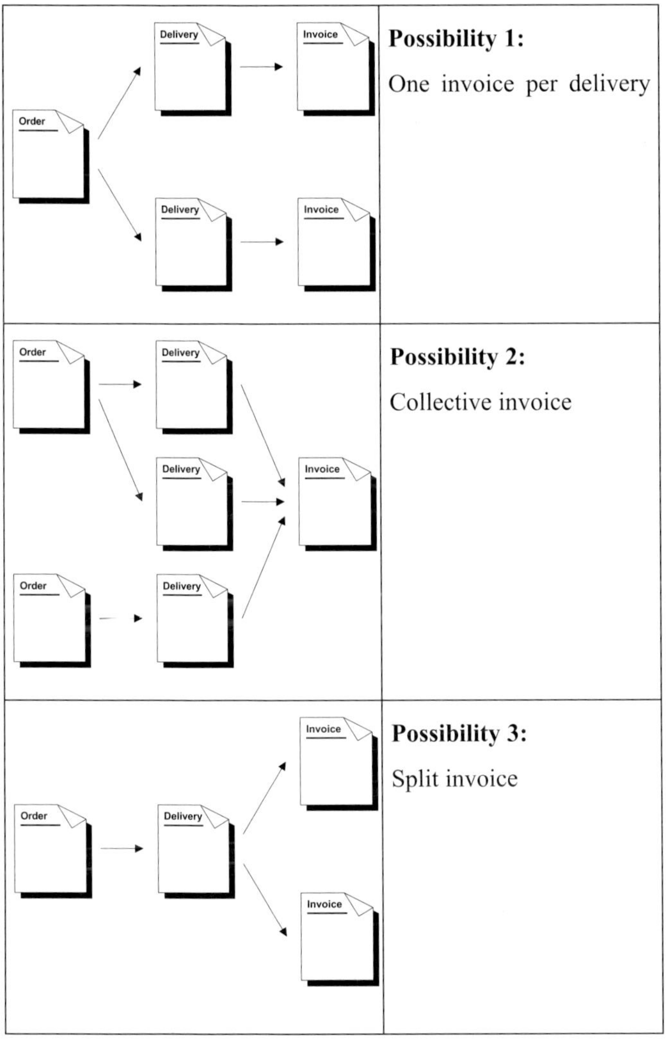

Figure 3-43: Splitting and combining of invoices

SAP Retail also permits the generation of *invoice lists*. These are often used when different customers have the single payer. This, for example, is the case for purchasing cooperatives where all invoices for goods deliveries to the cooperative members are paid centrally. In a two-step procedure, first the billing documents are processed as in a normal run. However, an additional lock flag is set for the payment and dunning. Thus, payments are neither expected nor requested, similarly no reminder notices are produced when the payment deadline is exceeded. In the second step, an invoice list is produced, sent to the payer, and the lock for payments and reminder notices is cleared.

The data are available in the other SAP modules as soon as the billing has been performed. The posting records are transferred immediately to the accounting module. This allows all FI applications and the analyses of the controlling to access the new data in real time.

Creation of credit notes and debit memos

A posting correction can be used to correct billing documents. Here, all postings that resulted from the invoice are revoked with the corresponding offsetting postings. If only parts of an invoice need to be corrected as result of a return, credit notes or debit memos are created for the associated return delivery. After release, the system automatically makes the necessary payments to the customers.

Figure 3-44 summarizes the sequence of the documents in the distribution process. If an order invoicing is performed, the delivery can be bypassed. The gray areas (materials management and financial accounting) indicate other important modules, with which sales and billing exchange data.

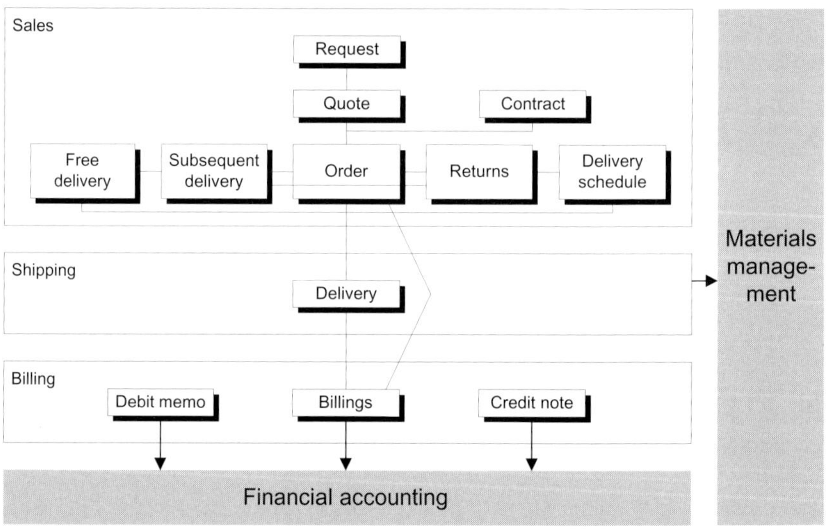

Figure 3-44: Document flow in sales and billing

4 Distributed Retail Information Systems (DRS)

4.1 Business Implication

The specifics of the business performed by retailing companies normally require territorially distributed structures. Thus, a retailing group usually consists of central areas, regional subsidiaries, central warehouses or goods distribution centers, and stores. In addition to the local distribution, the various retailing companies have very different organizational structures. Similarly, the distribution of the functions is often greatly influenced by the industry and the company philosophy. Consequently, it is hardly possible to provide a general distribution model for retailing limited to just a few scenarios. For these reasons, a modern retail information system should provide a large degree of flexibility to allow it to be adapted to any distribution structures.

Additional requirements made on the distribution result from the growing customer-orientation, flexibility, and the trend to increased decentralization of competencies. Central coordination and decentralized handling should be integrated. Centralized retailing management is increasingly being augmented with local or regional-oriented micro- or segment-marketing. The resulting distribution of the functions to geographically separated organizational units requires the distribution of the retail information systems that support these functions.

Distributed retail information systems (DRS) can be used to perform functions decentrally. Not only vendors and customers, but also mobile field sales representatives communicate with such systems. A connection to the Internet is assumed nowadays. However, because of the historical development of data processing in many retailing companies, heterogeneous DP landscapes often exist with various subsystems. In many cases, such subsystems are either not integrated or are poorly integrated (Hertel 1999). This makes it difficult to satisfy current requirements, let alone future requirements.

State-of-the-art for the realization of DRS in retailing companies involves the use of integrated standard software that is tailored to the industry and which can be customized to the requirements of the individual company. Such standard software should have defined interfaces based on recognized standards for the distribution, and for the integration of external and partner systems. In particular, the last aspect permits the distribution of the functions to different systems often needed in practical use.

A prerequisite for the realization of distributed systems with integrated standard software is the capability to represent the generally relatively complex legal and organizational structure of the associated retailing company. This is provided in SAP Retail using R/3 standard structures and additional retail-specific supplementary structures that offer adequate flexibility (cf. Section 3.1). An example is shown in Figure 4-1.

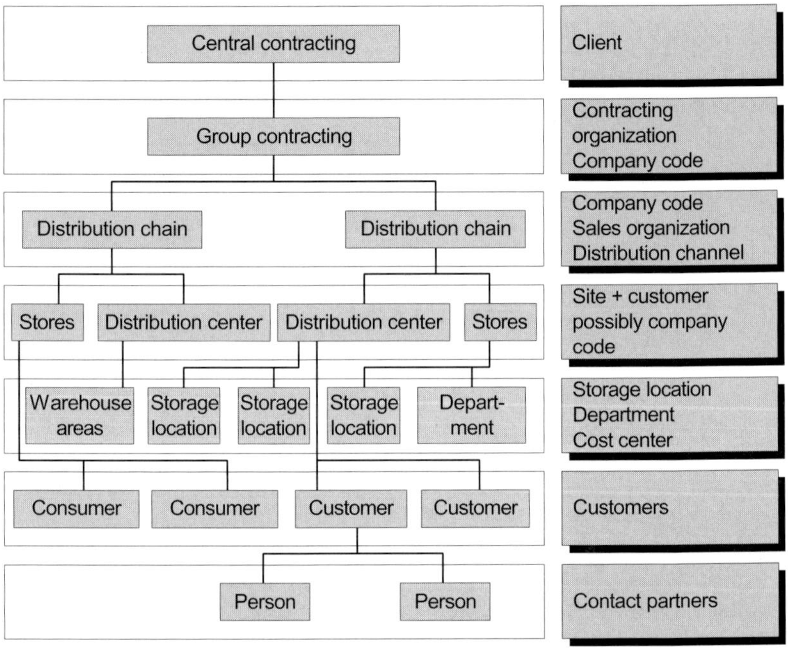

Figure 4-1: Sample representation of a typical retailing structure in SAP Retail

The primary business significance of the distribution is that it permits parts of a company to operate independently. For example, stores can make their own requirements planning and transfer their store orders to the head office, which then only consolidates these to large purchase orders and so can optimize the ordering quantities at the vendor. In addition, because of the very large volumes of master and transaction data processed in retailing company, the performance advantage permitted by distribution should also be emphasized. Even a loose coupling of the systems that support the logistics with the systems for human resources that can be realized with distribution, increases, for example, the flexibility of the complete application system. The integration with application systems of vendors and other business partners based on distribution permits the optimization of inter-company business processes.

In this chapter, we initially use an example scenario to show the organizational and business relevance of the distribution of application systems in retailing and also provide an overview of the technological capabilities of the realization of distributed systems based on SAP Retail. The following sections discuss these capabilities in more detail.

4.2 Example Scenario of a DRS

For the reasons explained in the previous section, we describe here a typical example rather than various possible concrete distribution scenarios. This example illustrates the most important capabilities of the territorial and functional distribution in a retailing company. A retailing group with stores that acts both as wholesaler and retailer for electronic components and hobby articles in the market is used to present the various technologies available for the distribution.

Figure 4-2 shows an overview of a possible distribution structure of the application systems for this company using SAP Retail as basis.

Structure and function distribution

The nationally operating group has a head office that handles the procurement, personal management, controlling and the complete financial accounting. In the goods distribution centers, large deliveries (direct supply from the manufacturers) are accepted, divided and forwarded to the stores. For this purpose, a special logistics system purchased and installed prior to the implementation of SAP Retail should continue to be used in parallel to SAP Retail.

The company uses three channels to market its articles:

1. The group has electronics department stores (designated as stores here). In some of these, customers have the possibility, in addition to the usual buying, to use POI (Point-of-Information) terminals to obtain information about the current availability and to order certain goods for home delivery . The store employees can also retrieve data about articles, vendors, customers, purchases and orders and, to some extent, update these. For example, the sales assistants can initiate purchase orders and so respond flexibly to requirements.

2. Another distribution channel of the group concerns sales to bulk buyers that field sales representatives handle. As with the store employees, the sales representatives can retrieve and update data for customers, articles, etc.
3. The third distribution channel of the retailing group involves sales over the Internet. These can take place at self-service terminals in the stores or by using Internet providers of the customers.

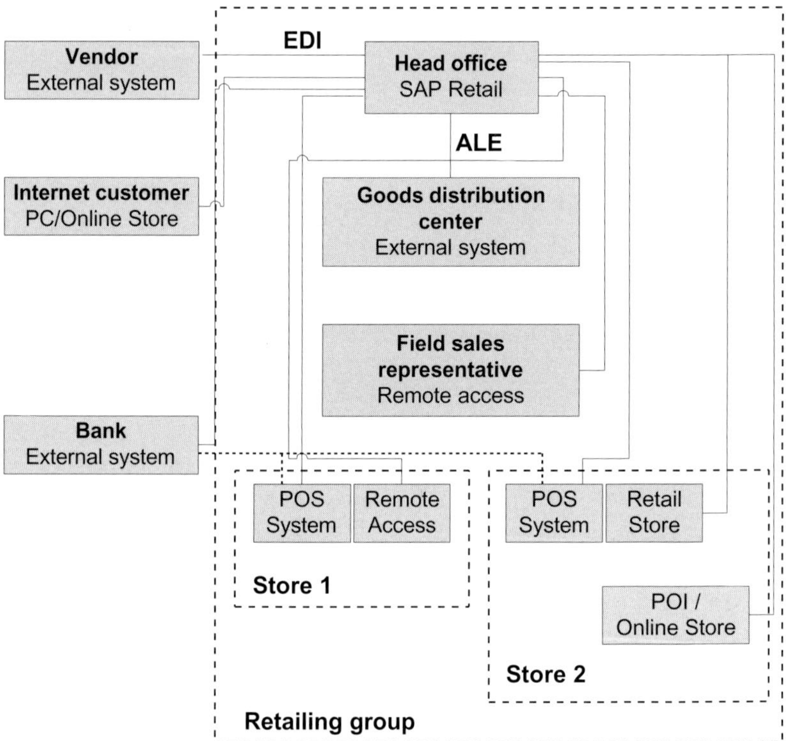

Figure 4-2: Example of a distribution scenario using SAP Retail

Table 4-1 shows some of the important data types needed for the realization of the functions, assigned after the creation, the maintenance and the use in the scenario shown above.

Table 4-1: Possible distribution of some data types in the example scenario

Data type	Source / Creation	Maintenance	Analysis / Use
Article data	head office	head office	head office, store, field sales
Vendor data	head office	head office	head office, store, distribution center
Customer data	field sales, store	head office, store, field sales	head office, store, field sales
Personnel data	head office	head office	head office
Personnel job data	store, distribution center, field sales, head office	head office	head office
Sales order data (requirements planning)	head office, store, field sales	head office, store, distribution center, field sales	head office, store, field sales
Good receipt data	distribution center, store	head office	head office
Sales data	store, field sales	head office	head office, store

Technical realization of the distribution

The technical realization of the distribution in the example scenario could have the following form:

The head office has the SAP Retail system on appropriately powerful hardware. A special (non-R/3) system is used to optimize the logistics in the goods distribution centers. This special system communicates over the R/3-specific ALE (Application Link Enabling) interface with the SAP system. The company-internal intranet (firm network with Internet technology) is used as transmission medium. In order to perform quickly the frequent orders with the vendors, these are transferred using EDI (Electronic Data Interchange). The three distribution channels are to be connected to the system in the following manner:

1. The stores have remote access to the Retail system and so can use the complete stored data and functionality. The access is performed encrypted over the public telephone/ISDN network. There still exist POS terminals with which the employees at the cash desks can access the article master data of the Retail system and book sales. There is no continuous online connection between Retail system and point-of-sale system. Instead, the changed master data (in particular, prices) are normally transferred during the night to the affected stores. The stores send their sales data, which the SAP system subsequently processes. The requirements planning and the purchasing are then performed at the head office using the updated data. A capability for a connection to the bank is also provided at the POS. In this case, the electronic payment traffic can be handled using the Retail system. Compared with the remote access, the use of SAP Retail Store provides a much more modern and easily adaptable means to the access the Retail system. This component offers a user interface tailored for

handling the store-oriented processes. The interface focuses on functions and data required for the store retailing, hides complexity from the user and so can be operated easily by less-experienced personnel. SAP Retail Store or remote access can be used to directly enter and monitor sales orders independent of the POS system. Queries, quotations, scheduling agreements, billing documents, etc., can also be handled in the same way. These options allow the stores to plan independently. The created store orders are then grouped at the head office to larger orders with correspondingly optimized order quantities.

2. The sales to bulk buyers are handled by the field sales representatives, who, similar to the employees in the stores, have remote access to the complete SAP Retail functionality. A modem or ISDN card connected to the telephone network or a mobile telephone connected to the wireless network can be used to transfer data with the SAP Client or a Web browser. SAP also supplies special software with the name SAP SALES that is tailored to the requirements of the field sales and the efficient customer support. Section 7.2.5 discusses this product that is part of the "customer relationship management" New Dimension Line.

3. The SAP Online Store component permits sales over the Internet. This is done with the intranet at the self-service terminals installed in the stores or directly using the Internet on the customer's PC.

The following sections introduce the most important capabilities of the distribution in more detail.

4.3 Remote Access

Remote access offers the technically most simple means of distribution or access from remote locations to SAP Retail. This uses an SAP front-end to access the data base and the functionality of the Retail system. Possible scenarios for the use of the remote access are, for example:

- An employee in a store enters the results of a physical inventory in the system.
- A vendor views current sales data at the retailing company.
- A store manager retrieves at home data about the current sales.
- A field sales representative retrieves the current data and the availability for specific articles or enters current orders.

Access is normally made over a dial-up line or leased line using a modem or an ISDN card. Obviously, access using a mobile telephone with data transfer option is also possible. Because of the rapidly increasing availability of Internet/intranet, its use for access is gaining in importance; all front-end types can be used for this purpose. The TCP/IP protocol allows the SAP GUI to use the Internet to access the server on which SAP Retail is installed without needing to make any modifications. Current versions of SAP GUI can be used with older R/3 releases (downward-compatibility). The following sections provide a short description of the three current types of SAP front-ends:

SAP GUI for the Windows environment

This front-end, previously known as SAPGUI (SAP Graphical User Interface), has been available since the beginning of the R/3 development but previously was almost always used exclusively for internal purposes. It is suitable for experienced SAP Retail users who use the system frequently. It is available on the following platforms :

- Windows 95 with restrictions
- Windows 98, Me and Windows NT 4.0 (only the Intel platform)
- Windows 2000 (Ver. 4.6D or higher recommended)

SAP GUI for the Java environment

From Version 4.6D, a Java front-end is available for the platforms

- IBM OS/2
- Apple Macintosh
- LINUX
- Various UNIX platforms (AIX, HP-UX, Solaris, etc.)
- A Windows version is also available (for use within SAP Workplace)

Both introduced types of SAP GUI are envisaged for frequent users of the R/3 system. The function elements have been kept largely platform-independent. Figure 4-3 shows the new “look & feel” user interface of this client that is better tailored to the user’s needs (EnjoySAP).

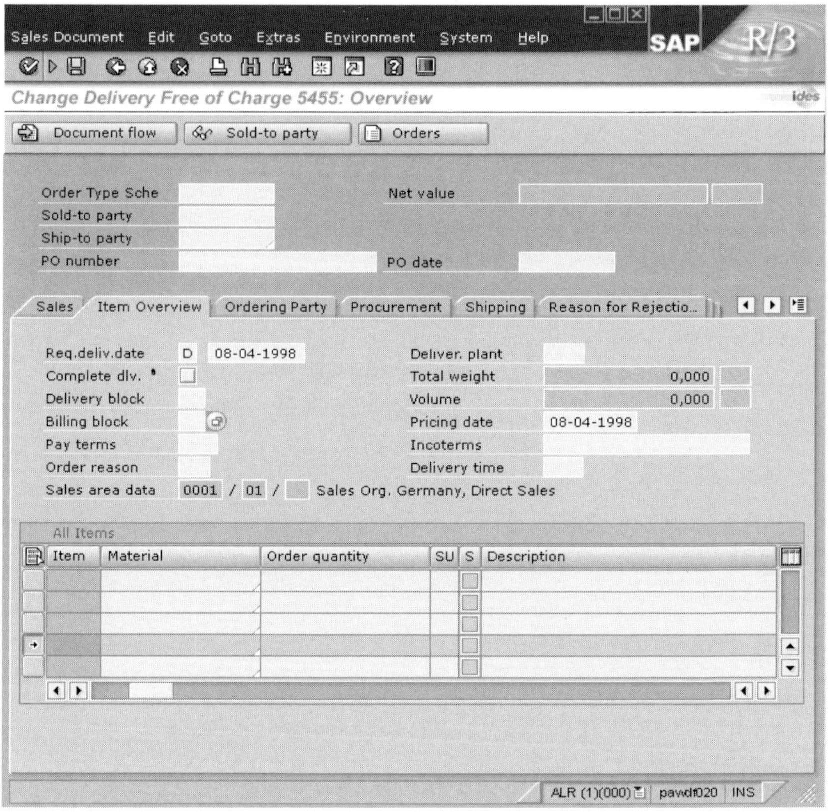

Figure 4-3: New user interface of the SAP GUI (from Version 4.6)

Internet/intranet applications / SAP GUI for HTML

The use of Web-based portals is currently rapidly gaining in importance. The reasons here are the wide availability of the technologies, the ease of use and the high acceptance of this medium. The primary goal is the development of a user interface with limited functionality tailored to the occasional user. So-called IACs (Internet Application Components) available for a wide range of application areas allow the use of these selected functions. SAP with R/3 Release 3.1 for the first time has expanded the supplied components. An important application example is introduced in Section 4.5 with the SAP Online Store. To use the IACs, an appropriate adapter element to the SAP system is needed in addition to a Web server. This component, designated as ITS (Internet Transaction Server), is supplied with the SAP system. Chapter 6 contains a more detailed explanation.

There are further possibilities to connect the Retail system to the Internet, e.g., using programs developed in-house or using third-party products and the SAP Automation program interface. Because the scope of this book does not permit these to be discussed here, we recommend Perez 1999 and the SAP Online documentation (SAP library) as sources for further details.

Table 4-2: Comparison of the front-end types

	No local administration	Ad hoc use	Total functionality	Front-end components
SAP GUI for HTML / IAC	✓	✓		
SAP GUI for the Java environment			✓	✓
SAP GUI for the Windows environment			✓	✓

Summary

In general, all three front-end types are connected to SAP Retail using the Internet/intranet. However, the SAP GUI for the Windows and Java environment uses the Internet only as transport medium, whereas the SAP GUI for HTML and the Internet/intranet applications with the Web browser can use additional Internet technologies, e.g., the World Wide Web (WWW) service and the HTML document description language. Whereas the SAP GUI for the windows and Java environment form an independent R/3 client program, SAP GUI for HTML needs only a browser at the client computer and does not require any local administration. Table 4-2 compares the most important properties of the front-ends.

4.4 SAP Retail Store

SAP Retail Store represents a user interface with additional typical store functions that can be customized directly to meet the stores' requirements. It is an independent software product that simplifies the use of SAP Retail by the personnel in the stores through the provision of specific data entry screens and supplementary functions, such as finding and automatically assigning organizational units.

The user interface is realized using the technology of the WWW Internet service, i.e., using HTML pages. This permits a simple, intuitive operation and the use of standard Web browsers. The use of HTML pages makes it easy to customize the user interface and the functionality. SAP supplies its own development tool, SAP@Web Studio, that can be used to modify the appropriately prepared IACs in the layout and functionality, even for people with limited HTML knowledge. Although SAP Retail Store runs as a logically independent system, it belongs physically to the (central) retail system.

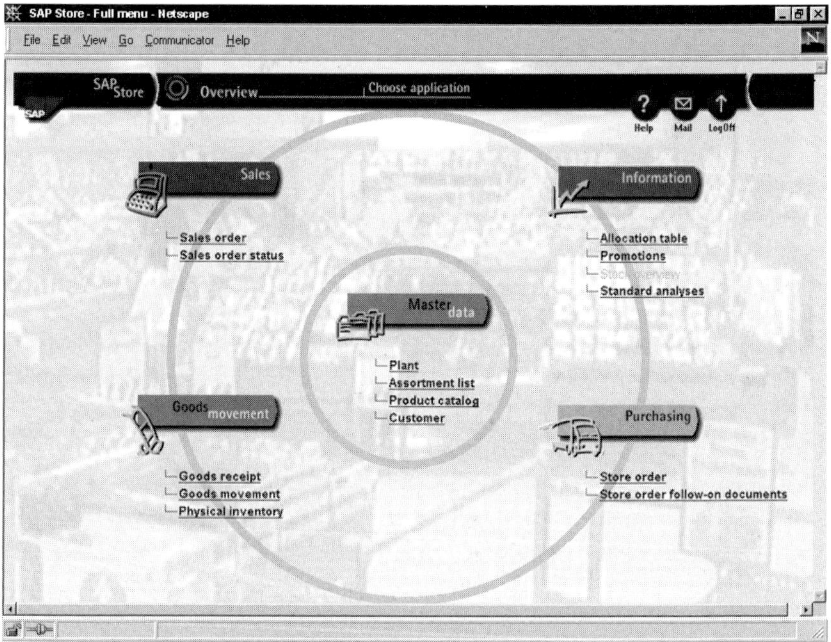

Figure 4-4: SAP Retail Store

The functions provided with SAP Retail Store support the storage of master data, contracting, goods movements, sales and information supply. They are used mainly for mass business and cover more than 90 percent of the business transactions. A direct access to the Retail system is available when the complete functionality is needed. Both access methods can be realized in the store using low-price standard PCs. Figure 4-4 shows a typical start menu for SAP Retail Store.

To better show the operational areas, some central functions of the SAP Retail Store follow:

- Display information about the assortment articles.
- Display the product catalog, namely, product information is prepared in multimedia form for the salesperson (with search functions, text search, etc.).
- Input, display or change of customer data.

- Direct Input of store orders. This function makes it possible to create the appropriate follow-on documents, such as order requisitions, purchase orders, deliveries or sales orders in the central Retail system. The system automatically determines the supply sources.
- Input of goods receipt (with references to the corresponding purchase orders).
- Input of stock transfers and physical inventories.
- Sales orders can be created in the system during the sales talks, in which price and availability are displayed. The associated status can be queried subsequently.
- There is also a facility to request various information and reports, e.g., inventory levels, order lists, sales and staff statistics and promotion reports.
- It is also possible to participate in e-mail communication using the SAP system (refer to Figure 4-5).

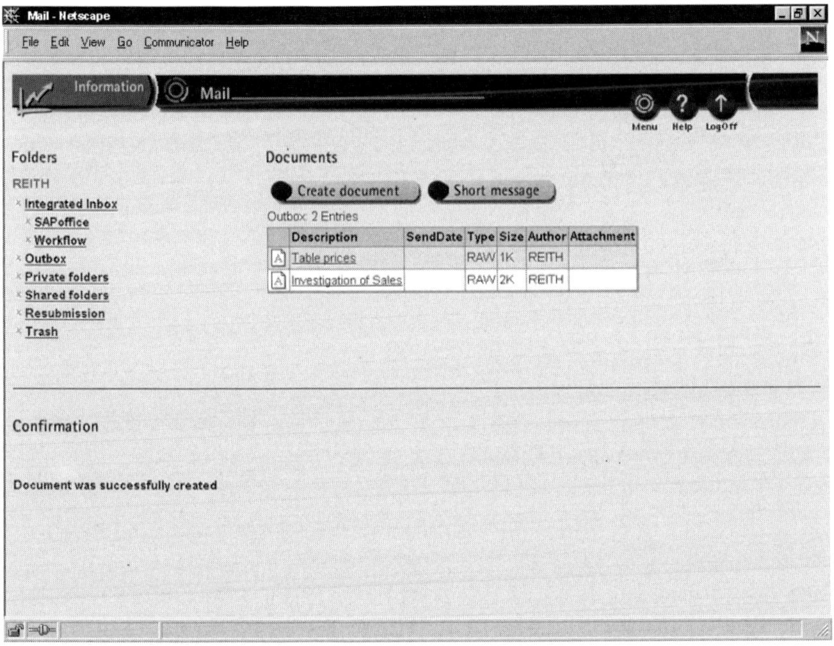

Figure 4-5: E-mail user interface in the SAP Retail Store

The SAP Retail Store functionality will be further expanded. To reduce the dependency on the central Retail system and to improve the performance, SAP will provide in new releases an additional mechanism for more autonomy of SAP Retail Store This will allow the use of functions in offline mode using special buffer technologies.

4.5 SAP Online Store

The SAP Online Store represents an Internet application component (IAC) of the Retail system that uses the base functions of the product catalog. It uses the SAP document management to provide multimedia objects such as text, graphics and photographs for the individual products. Thus, the Web applications operate directly with the data from the SAP system and do not require any duplicated data storage.

The Online Store provides the following positive effects in the business-to-consumer segment:

1. Because mail order companies do not need to print any catalogs (or at least fewer) and orders previously often made by telephone, letter or fax no longer need to be entered manually, there can be realized savings potentials.
2. Secondly, the Online Store permits the use of kiosk systems in the stores to give presentation support to the sales personnel for out-of-stock articles and information terminal for interested customers. The customer can also make orders there.
3. A third important aspect is the provision of new distribution channels for retailing companies. However, at the same time, a threat arises here when the producers directly market their own goods and so avoid the retailer as intermediary. This enables them to gather information about the customer behavior and thus achieve customer loyalty (cf. Chapter 6).

In the business-to-business segment, e.g., for purchases by retailers from wholesalers, synergies, cost savings and reduced error rates can be seen as positive effects resulting from the use of Online Stores.

The SAP Online Store forms the equivalent to SAP Retail Store from the customer's viewpoint. It also uses the Internet with the familiar standards and the user-friendly interface, and provides functionality tailored to the customer's requirements.

Figure 4-6 compares SAP Retail Store and SAP Online Store as components of a distributed system. The intermediate layer, called BAPI, for the access to SAP Retail is described briefly in Section 4.8.

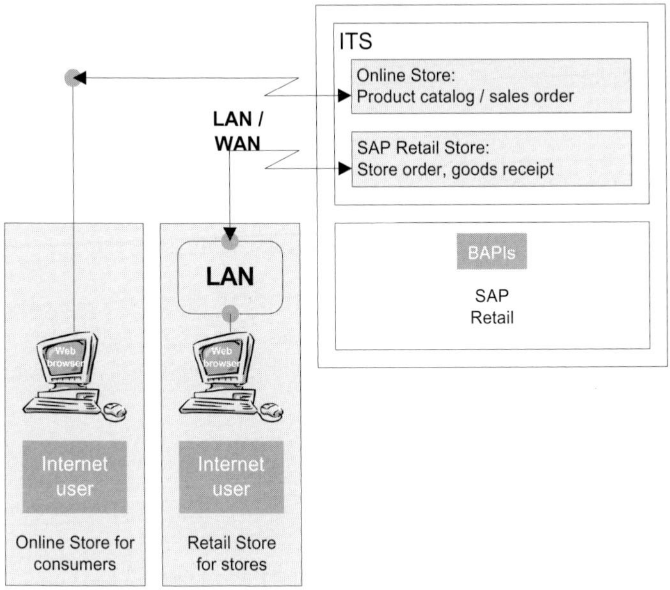

Figure 4-6: Comparison of SAP Retail Store and Online Store

The most important function of the SAP Online Store for the customer is fast a searching, an accurate description and a realistic display of the products. Search tools are provided for support. The presentation can be augmented with multimedia elements.

The Online Store provides other functions, such as the initial registration of new customers with the assignment of access data, a flexible shopping basket management and various forms of payment (including payment with credit card) with application-specific conditions depending on the customer, the merchandise category or other criteria. Two interfaces are provided for payment by credit card. The first is used when the SAP system requests the authorization from the clearing house. The second interface supports the SET (Secure Electronic Transaction) standard that is used for the secure processing of credit card payments over the Internet.

The system gives the customer immediate information about the availability of the desired article. Once the order has been made, the customer also has at any time the possibility to query the current status of the processing of the order.

The current version of the Online Store supports the linking with external Internet search engines, such as AltaVista, Fireball and Lycos, and also permits the use of SSL (Secure Socket Layer), a method to establish a secure connection between the Web browser and the Web server. Figure 4-7 shows an example of the Online Store user interface.

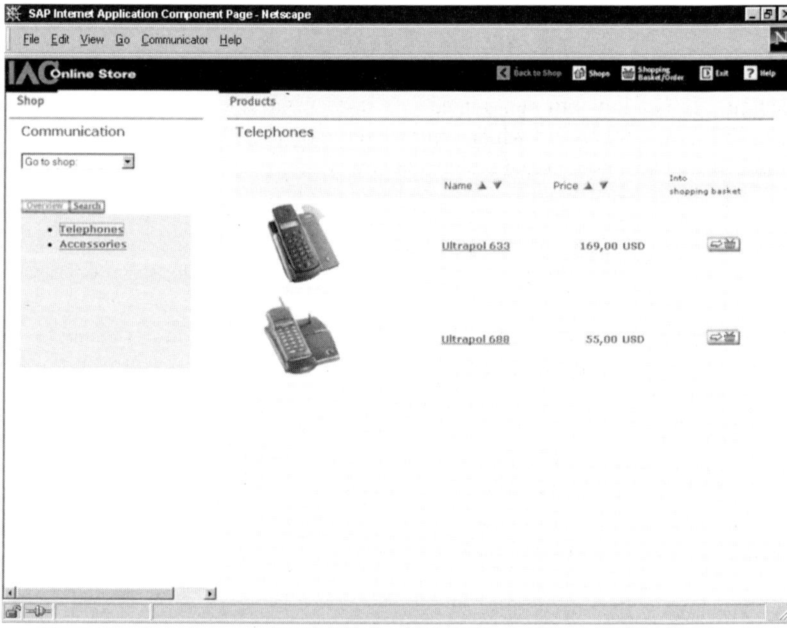

Figure 4-7: Example for an Online Store

The previously mentioned SAP@Web Studio development environment can be used to customize the Online Store to the requirements of the company.

4.6 Connection of External Systems using ALE

SAP with ALE (Application Link Enabling) offers a very powerful means of distribution. ALE supports the business controlled information exchange between loosely coupled applications while maintaining data consistancy. This technology permits the coupling of different R/3 systems, of R/3 systems and R/2 systems, and of R/3 systems and non-SAP systems. Its use requires the definition of a logical model for the complete system (distribution model) that specifies in which system which applications run and which information is exchanged between the systems. Asynchronous (and synchronous) communication rather than a central database is used for the application integration. Thus, distributed databases are not used here, but rather data files, managed redundantly in several databases. Examples for data exchanged between systems:

- control data (organizational units, customizing data, ...)
- master data (articles, vendors, customers, cost centers, ...)
- transaction data (invoice, shipping notification, purchase order, ...)

SAP has already prepared various distribution scenarios that can be used as sample solutions. For example, ALE scenarios are provided for the interaction with external warehouse management systems and for the distribution of contracting functions (separation into centralized and decentralized contracting). ALE integrates different systems but which remain autonomous in operation and updating. Messages are exchanged based on business transactions.

ALE provides good support to achieve the goal of a comprehensive flexibility for the distribution of data and functions within the retail information system of a retailing company mentioned at the beginning. Integrated information exchange over system limits allows a large degree of freedom in the system structure. This also affects the possibility of stepwise implementation of system components. For this and other reasons, a decoupling of the basis, logistics and personnel management of the Retail system is currently being realized using ALE. The associated positive effects are an improvement of the data security through the separation of the data files, the possibility to use different release levels in the mentioned areas, performance advantages, and the possibility to meet special distribution requirements of the companies. Because ALE can also be used to connect non-SAP systems and partner systems, it provides wide capabilities for the distribution of the retailing management in a heterogeneous system landscape.

Figure 4-8 illustrates the principle for the coupling of two systems using ALE.

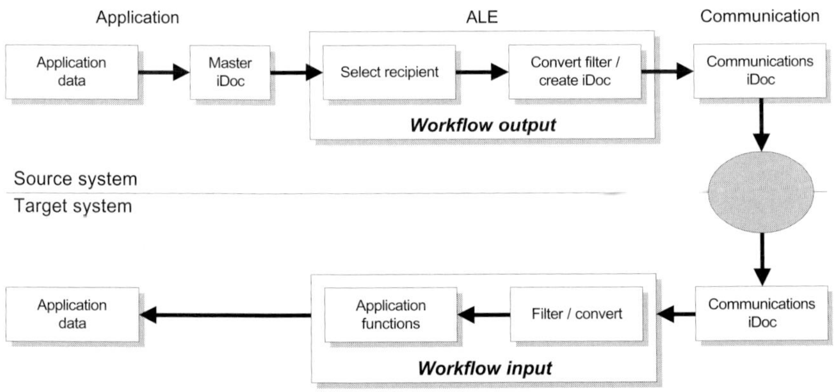

Figure 4-8: Coupling of two systems using ALE

The diagram shows how the processing can be divided into three sections: application, ALE and communication. The associated parameters for the smooth data exchange and the correct exception handling must be maintained in all three levels (Hernandez 1998, pages 1073-1077, Hernandez 2000). So-called Intermediate Documents (IDocs) serve as message container. They represent standardized document structures that are used to pass messages between systems or components. The so-called translator, which acts as intermediary when non-SAP systems are connected, converts the IDocs into the appropriate structures for the non-SAP system and controls the communication, for example, establishes the connection and performs a restart.

EDI (Electronic Data Interchange) is another important technology used in distributed retail information systems. It permits the automated exchange of business data in the business-to-business area. This interface is realized in SAP Retail using ALE on the basis of IDocs. Further discussions for ALE and EDI can be found in Kasturi (1999), Angeli (2001) and, in particular for the use of EDI, in Chapter 6 and in Wenzel (1999, pages 504-525).

4.7 POS Interface

Current business concepts and strategies, such as efficient consumer response, category management, customer loyalty and others (refer to Chapter 7 and Ahlert 1998, pages 343-398), require for their implementation a comprehensive recording of all sales and a complete integration of the cash desks or, generally, of POS (point-of-sale) systems in a distributed retail information system. Many marketing and management reports (see Chapter 5) for sales are not possible without exact article, time and cash desk data. The complete processing of payments, including the use of credit cards and/or customer cards, can be handled using the POS interface and the central financial accounting. A further advantage of the direct connection is the high actuality of the data for the articles and the price accuracy. This reduces the probability of making errors in sales. Figure 4-9 shows information flows originating at the POS.

Because of the large number of different POS systems, SAP Retail provides an universal interface that permits the connection to different POS systems. This allows both data to be retrieved from the Retail system, such as, article, customer and personnel master data and prices, and also data to be returned to the Retail system (e.g., goods movements, money transactions and statistical data). In addition, functions from SAP Retail, such as pricing, sales or payment handling, can be used.

Because no permanent connection needs to be maintained to the head office, the use of the POS interface gives the store a large degree of independence in the day-to-day business. The changed master data (e.g., prices) are normally brought up-to-date at night. Similarly, the day's sales and payment data are collected and sent together to the head office at the end of trading. The updated requirements planning, purchasing and order bundling can be performed there.

Because of the different requirements, the POS interface output and the POS interface input remain separate components. The POS interface monitor, which exists as third component, is used for error handling, etc. The following sections consider these three modules in more detail.

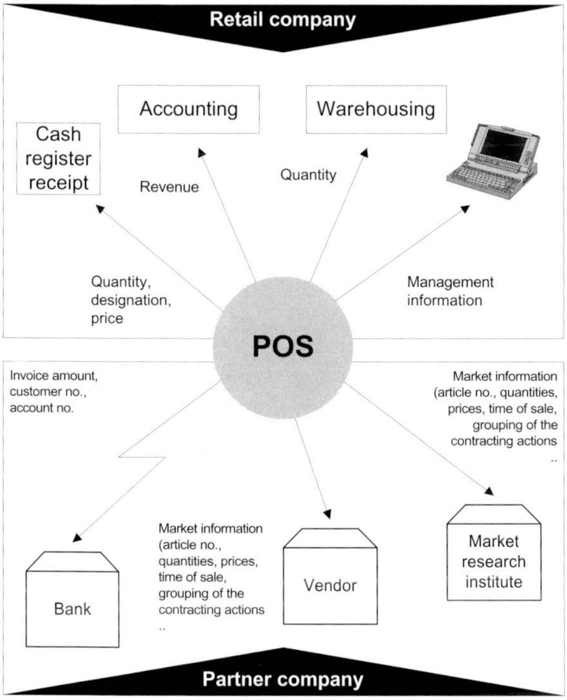

Figure 4-9: Information flows at the POS (Mertens et al. 1998, page 136)

POS interface output

The output is used to regularly supply the stores by download (head office to the store) with the current master data. The connection is performed using IDocs described previously in Section 4.6. Examples for the IDoc data structures needed here are merchandise categories, articles, EAN references, personnel data or currency rates. To customize the documents to the format required by the POS system, a special converter transforms structures stored by ALE in the IDoc database.

POS Interface Input

As with the POS interface output, the message shipping back to the Retail system (upload from store to the head office) is performed using IDocs. However, because of the various different functions available in the Retail system, the capabilities of the POS interface input are more comprehensive. Examples of data groups here are sales, payment lists, money transactions, goods movements, store goods receipts, cashier statistics, store orders, physical inventories, and master data updates. These examples make it obvious that this interface can also be used for data not produced directly at the POS terminal.

An example for a business process using the POS interface input follows:

- The scanner or the keyboard is used to enter the customer number, the article numbers (or EAN) and any discounts.
- The associated data, such as customer name, article name, prices and possibly other functions (e.g., calculation of special conditions for the customer or article), which the POS interface output regularly updates (normally daily) by download from the Retail system, are retrieved from the local POS controller.
- The data for the sales and the corresponding payment are collected in intervals and are transferred regular, e.g., at night, using the POS interface input to the Retail system for further processing.

As Figure 4-10 shows, the transferred data initiate functions of the Retail system in the inventory management and the financial accounting, payment transactions are forwarded to the appropriate bank, and the relevant data are made available in the Retail Information System. These data can be used not only for purchasing and contracting, but also serve as basis for management support systems (refer to Chapter 5).

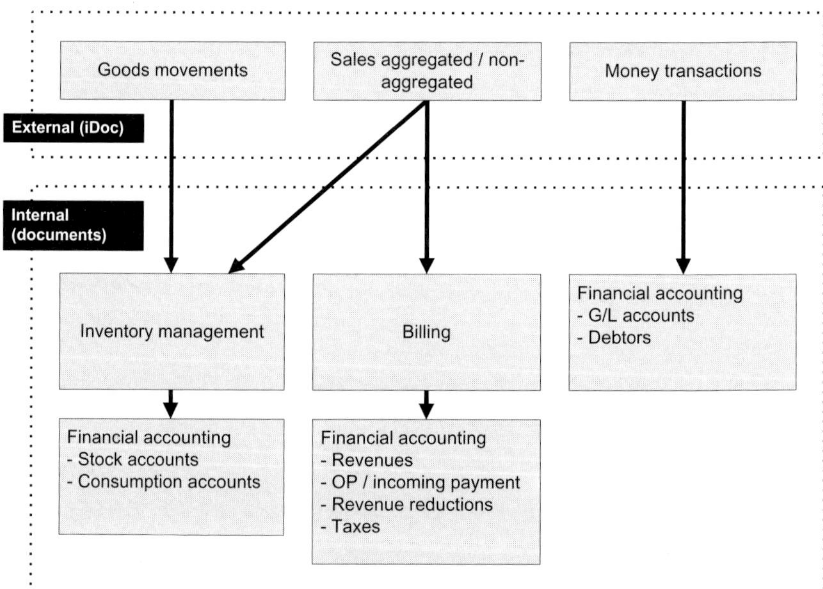

Figure 4-10: Transformation of the IDocs in documents for the POS interface input

POS interface monitor

The POS interface monitor is used to monitor the POS interface output and the POS interface input. It does this by classifying and logging all output and input messages. It is possible to represent all processed messages using codes for the various criteria and including any follow-on documents. Messages from point-of-sale systems can be displayed. Postings can be processed and cancelled. The processing can also be re-initiated.

Should problems arise with the IDoc processing, the POS interface monitor is capable of initiating the appropriate processes for error handling. A differentiation is made here between errors and exceptions.

- Errors result from incomplete IDoc data, incorrect customizing data for the definition of the communications paths or incompletely updated master data (e.g., for the assignment of article master data). As response, the POS interface monitor permits the reprocessing of the transmission data.
- Exceptions are special IDocs that must be reworked or corrected, and then reprocessed. It is also possible to forward them to specific processing clerks. The workflow components from SAP R/3 can be used here to maintain corresponding roles for certain exception messages. Criteria can be specified for the assignment. Figure 4-11 illustrates the described procedures.

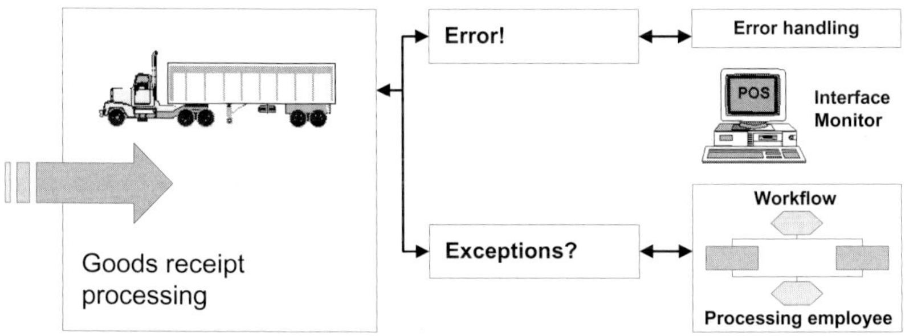

Figure 4-11: Operation of the POS interface monitor

POS converter

As mentioned previously, a conversion of the data structures is needed to connect the different point-of-sale systems. The POS converter provides this function. It is not part of the SAP Retail system, but is offered by SAP partner companies.

Figure 4-12 shows the required IT structure to connect point-of-sale systems to the Retail system using the POS converter. As apparent in the figure, the IDocs destined for the POS systems are sent via the head office LAN to the POS converters. These are installed on dedicated hardware and may be duplicated to improve performance and increase reliability.

The point-of-sale systems of the stores are coupled using a WAN in the form of modem or ISDN connections or using the Internet. The connection is established only when necessary. The data comparison is performed in regular intervals, normally daily. Each point-of-sale system has a POS controller that the individual electronic cash registers access. The server performs the communication with the head office and retains the required data locally.

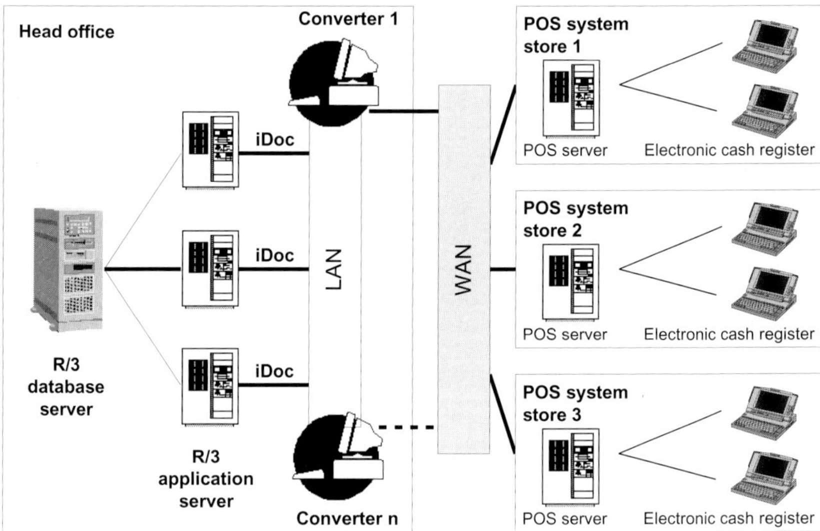

Figure 4-12: IT structure for the connection of POS systems to SAP Retail

4.8 Methodology Summary of Interfaces and Technologies

As shown in the previous sections, there exist various means, mainly relatively easy to handle, to distribute a retail information system using SAP Retail. These make obvious that SAP provides various open interface standards to connect non-SAP and partner systems.

Table 4-3 summarizes the characteristics of the distribution capabilities. This considers only the most important technologies discussed in this chapter. Some of these make use of each other.

Depending on the operational goal, the project team responsible for the distribution can make use of various possible interfaces. The remote access with SAP Retail Store and the Online Store has a direct connection to SAP Retail; thus, these capabilities – with the exception of the Store Engine – can be used only when access exists to the SAP Retail system. Message-oriented interfaces, such as ALE or EDI, can be used to connect autonomous external systems at the same level; these interfaces themselves require remote function call (RFC) or the EDI subsystem in SAP Retail to transport the data.

Table 4-3: Classification of the most important distribution technologies

Distributions capability	**SAP-specific technologies**	**General technologies**	**Usable networks**
Remote access	SAP GUI for the Windows or Java environment, IAC/ITS	ISDN adapter/ modem, Internet standards	Telephone network / ISDN, Internet
POS connection	ALE, Idoc	Limited Internet standards	Internet/intranet, LAN, WAN
SAP RETAIL STORE	IAC/IST	Internet standards	Internet/intranet, (LAN, WAN)
SAP ONLINE STORE	IAC/IST	Internet standards	Internet
Connection of external systems	ALE, IDoc, BAPI, RFC	EDI	LAN, WAN, Internet/intranet
Data exchange using standard interfaces	ABAP, BAPI, BDC	SQL	LAN

Function-oriented interfaces in the form of BAPIs (Business Application Programming Interface) are also available to connect external systems and data exchange using SAP standard interfaces. The BAPIs are business interface objects that permit precisely defined read and write access to the SAP data over individual release levels. Whereas the ALE technology (refer to Section 4.6) realizes the semantic synchronization of the data exchange, the BAPIs form the lower-level function-oriented business interfaces. BAPIs can also be used from other processing or communications components. In the meantime, several hundred BAPIs are supplied in the current R/3 releases. Moser (1999) provides a detailed discussion of BAPIs and their use.

SAP uses BAPIs to offer other interfaces in addition to the POS interface, such as for shelf optimization systems, personnel planning systems (Campbell) and to the SAP New Dimension Products (e.g., APO, BW) (also refer to Chapter 7).

Because the data exchange using standard interfaces provides only a relatively loose coupling with data matching normally being performed just daily or even more seldom, the data-oriented interfaces SQL (general standard for reading and writing databases) and BDC (Batch Data Communication, SAP standard for automated data transfer) can be used together with the ABAP programming language. Readers interested in further information about data exchange using standard interfaces should consult more advanced literature, such as Mende (2000) and Kretschmer (1996).

5 Systems for the Support of the Company Management

5.1 Problem Situation

One factor that particularly affects the competitiveness of a retailing company in the market is the availability of current management-relevant information supported by an adequate information and communications infrastructure. Information is a critical success factor, and, nowadays, business success results primarily from an information lead over the competitors. Executives need precise information from all areas of the company and from its environment. The large volume of operative data in retail information systems can not be considered as suitable as basis for making management decisions without sensible conditioning and possibly coupling of these data with information from external sources. Managers must be able to recognize any problem areas and to analyze their creation as fast as possible through the availability of meaningful key figures. The same key figures should be used for the support planning. Special *management support systems* (MSS) are needed here.

MSS are information and communications systems to support management. Their functionality is oriented towards the specific requirements of their users. MSS form a relatively independent subsystem as part of the information and communications system of a company. They provide computer services with which the manager is supported individually in his own organizational limits. He uses the computer service to improve the effectiveness and efficiency in performing his tasks (Frackmann 1996). MSS are innovative applications for modern information and communications technologies that are more similar to the nerve system in a living organism than to a conventional operative application system. They are used for the *information supply* (summarization, selection, linkage and presentation of management-relevant information), the *decision support* (application of methods and models for analyzes, forecasts and planning), the *communications support* (integration of electronic media) and the development or verification of the "internal model" of the user, namely, they support his learning process.

Specific characteristics of management support systems are:

- the automated consolidating of data from various internal and external information sources
- a data base separated from the operative application with the capability of fast and flexible access
- freely-selectable data selection and summarization; individual specification of information filters (exception reporting)
- a range of methods and functions for analyses, forecasts and planning, including early-recognition functionality, oriented towards the requirements of management
- a graphically oriented user interface; intuitive use; context-sensitive help
- simple change and extension capabilities.

Nowadays, MSS are normally realized using the data warehouse concept. A *data warehouse* is generally understood to be a company-wide data base for all forms of management support systems. The term was formed by Immon (1992). The basic idea of this concept is that the data are available to the user as in a warehouse. The user can retrieve and analyze these data using simple means. The characteristics of this concept are (Mucksch, et al. 1997):

- The data warehouse is supplied with data using defined interfaces in specific time intervals from the operative applications and from external sources. The import component used here also performs any required transformations, such as selection, aggregation, grouping or key conversion.
- The operative systems are strictly separated from the data warehouse.
- The data are stored user-oriented in the data warehouse, i.e., arranged according to functional areas / topics.
- The data relate to a specific period or time; they are provided with a so-called “time-stamp”.
- The data warehouse is fully integrated (semantics: designations, definitions etc.; logical: data formats).
- Characteristic is the non-volatility: no further changes are made to the data in the data warehouse.
- A meta-information system is used for the description and administration of the data in the data warehouse.

- The user can access both the original data and the prepared or summarized data.
- The user is provided with a powerful standardized user interface, and defined access and analysis methods. These methods range from standard reports, include ad-hoc queries, through to sophisticated data analysis methods, as grouped under OLAP (On-Line Analytical Processing) and data mining. The user can normally import the results of the analysis into various front-end tools in his office environment, such as worksheets, and can further process these data there, for example, for graphical presentations.

A major function of MSS in retailing is to obtain meaningful key figures from an almost endless volume of individual data that arise during the course of the business processes in the retailing company. Traditional success key figures in retailing are (Barth 1996, Falk, et al., 1992):

- *sales* in quantity and value
- *employee productivity* = sales (value) / number of employees
- *area productivity* = sales (value) / presentation area
- *shelf productivity* = sales (value) / contact distance
- *average inventory level*
- *area intensity* = average inventory level / presentation area
- *inventory turnover* = sales / average inventory level as productivity measure for the goods usage
- *range of coverage* = current inventory / average daily sales
- *various cost key figures*, in particular, using the activity-based costing
- *gross yield* (profit margin I) = sales (net sales price) – goods usage (net purchasing price)
- *gross margin* (as discount margin) = gross yield * 100 / sales price
- *gross margin* (as markup margin) = gross yield * 100 / purchase price
- *direct product profit* (profit margin II) = gross yield – directly assignable handling costs (Günther 1993)
- *various profitability* and *financing costs key figures*

In addition to these key figures, the industry-specific analyses characterized in more detail in Section 5.2.4 are typical for the management in retailing. Each of these analyses is itself based on key figures that need to be obtained from raw data and each produces specific more complex key figures. For example, *market-basket analyses* (Fischer 1997) require the following key figures which all refer to the analysis period:

- sales (quantity): sales quantity in quantity units
- sales (value): sales value in the local currency
- number of sales documents at the POS (cash register receipts)
- number of items (articles) on the cash register receipt
- number of items on the cash register receipt that apply to a promotion
- number of merchandise categories contained on the cash register receipt
- number of customers for which customer master data are available
- number of anonymous customers who paid with a credit card for which no customer master data are available
- number of customers without identification
- average price = sales (value) / sales (quantity)
- average number of items = number of cash register receipt items / number of cash register receipts
- ratio of promotion purchases = number of promotion items / number of items
- average cash register receipt sales = sales (value) / number of cash register receipts

These base and calculated key figures are obtained from the POS scanner data and must be available in the data warehouse. These key figures can themselves be used to calculate more complex key figures. An example here is the *conjoint profit* that is an expression of the profit contribution of an article in the total assortment (Recht, et al. 1997). When

- $DBWK_i$ profit margin of the shopping basket i
- DB_k profit margin of the article k
- x_{ik} quantity of the article k in shopping basket i
- p_k net sales price of the article k
- λ_{ik} ratio of the article k to the sales of the shopping basket i

then $$DBWK_i = \sum_k x_{ik} * DB_k$$

and $$\lambda_{ik} = (x_{ik} * p_k) / (\sum_k x_{ik} * p_k)$$

The profit contribution DB_{ik} of an article k in the shopping basket i then simplifies to

$$DB_{ik} = DBWK_i * \lambda_{ik}$$

and the conjoint profit CP_k of the article k to

$$CP_k = \sum_i DB_{ik}$$

The conjoint profit considers an article not in isolation but takes account of the companion effects within the assortment.

The article and its groupings, such as merchandise categories at different hierarchy levels or assortments with their modules, are a type of management-relevant classes of data objects. Further such classes are vendors, sales types, customers and their groupings, employees, presentation areas, contact distances, promotions, shopping baskets, stores, regions or periods. These classes and their objects must be identified and be made available in the MSS as reference quantities for key figures. Every sensible combination of such classes (e.g., customers, merchandise categories, regions and months) generally covers an n-dimensional data space in a data warehouse that must be filled with key figures (e.g., sales).

5.2 The Retail Information System (RIS)

5.2.1 Assignment

Generally, there are two options in the SAP world to build information systems to support the management in retail companies.

The integrated functional management information systems, which are components of the *Logistics Information System (LIS)*, can be used in SAP Retail. Although these information systems – the sales information system and the purchasing information system, the inventory controlling, the quality management information system and the *SAP Retail Information System* (RIS) – are built in modular form, they provide many "central methods" to analyze the data (method integration). However, the modular form also permits specific characteristics for each information system. The logistics information system at the user department level is part of the Open Information Warehouse (OIW). Further systems, such as the *Financial Information System* (FIS) or the *Personnel Information System* (PIS), are also available at this level. The Executive Information System (EIS), as component of the company controlling, can be used across user department boundaries (see Figure 5-1).

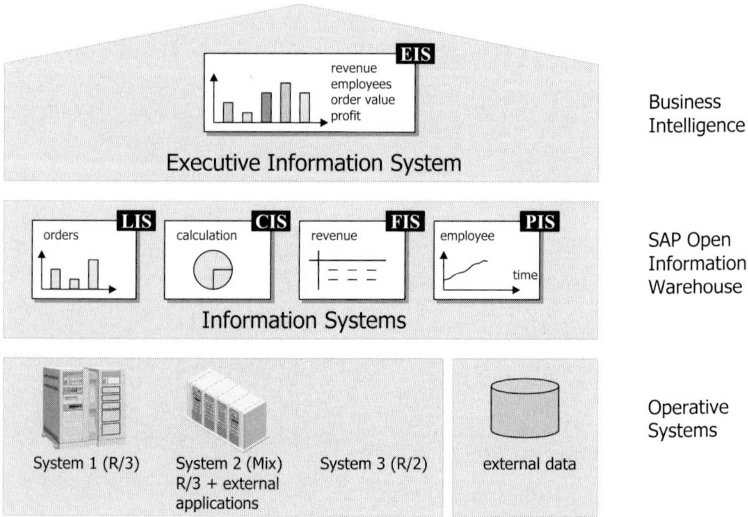

Figure 5-1: Management support systems in SAP Retail

The OIW is more a summary of the terms for the individual management support information systems. The historically developed information systems of logistics (LIS), of finance (FIS) and personnel (PIS) do not have a homogeneous structure and realize the basic principles of the data warehouse to different degrees. The OIW catalog serves as meta-information system. Microsoft EXCEL is used primarily as the front-end tool.

A new option for the structure of an MSS results from the use of the *Business Information Warehouse (BW)* provided as part of the New Dimension Products from SAP. SAP BW is a component that is independent of SAP Retail and must be coupled with SAP Retail and other components as part of the business framework architecture. Data sources can be the Retail system, non-SAP systems, and external systems. Because of its better technical infrastructure, SAP BW from release 2.0A is recommended when the use of SAP Retail is planned. Section 5.3.2 provides a detailed description of the advantages of SAP BW compared with RIS/LIS.

The following section first shows how information systems integrated in SAP Retail can be used. The discussion focuses on the retail-specific RIS that can be used to collect, summarize and analyze data for the retail handling, in particular. The section ends with a discussion of the SAP BW.

5.2.2 Data Basis

Characteristic for the information systems of the LIS is a separate data basis that is permanently supplied with data from the operative applications (sales, purchasing, financial accounting, etc.), and which permits a fast and user-friendly access. This data basis is formed using so-called *information structures*. An information structure defines a group of fields and is used to collect and aggregate data from the operative applications.

The data aggregation is necessary in an information system in order to obtain decision-relevant information. The aim is to recognize important interactions and trends from the many individual data items of the operative processing. Because many different user groups determine the requirements for data aggregation, a rigid specification of the aggregation does not suffice. Different user groups require different views of the data aggregation. Experience has shown that the standard information structures supplied with SAP Retail cover many typical requirements. Furthermore, the information from the operative applications can be selected within user-defined information structures that are considered as being important for the aggregation from the specific user viewpoint but that are not contained in the standard information structures.

An information structure consists of three types of information (see Figure 5-2):

- **Characteristics** are classes of management-relevant data objects or grouping criteria in the data warehouse, such as store, region, article, merchandise category or customer. Characteristics have characteristic values/forms that represent the individual objects of these classes (e.g., region: Saxony, Thuringia, etc.). The characteristics can be used to build hierarchies along which aggregations can be made (e.g., merchandise category hierarchies); characteristics can be independent of each other (e.g., article, region) or dependent on each other (e.g., article, merchandise category).

- **Key figures** are quantities, such as sales, order quantity or lead time, whose values related to characteristic forms or combinations of characteristic forms provide management-relevant information. They can be aggregated along the management-related hierarchies (e.g., merchandise categories). Company-specific key figures can also be described using formulas in user-defined information structures.

- **Period split**: the periodic accumulation of data is another aggregation criterion. Here, the key figures are summarized daily, weekly, monthly or depending on a variable period split determined using the business year variant.

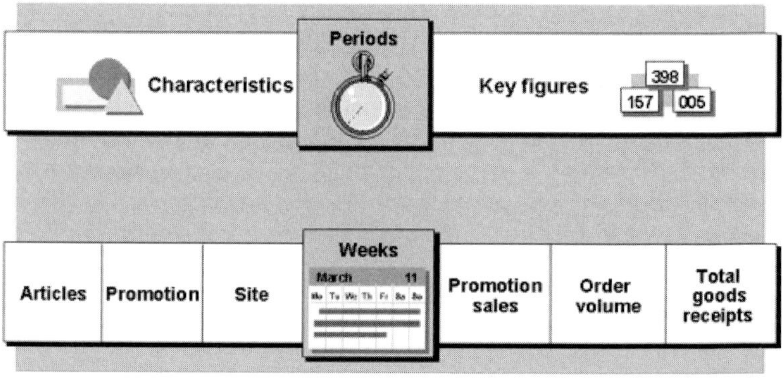

Figure 5-2: Sample representation of an information structure at a wholesaler

The n characteristics of an information structure cover an n-dimensional information space (Figure 5-3 shows an example with three dimensions). Key figures corresponding to the period split can be assigned to each combination of characteristic values. This information space forms the starting point for all analyses and planning to support the management of retail companies. Namely, the definition of characteristics, key figures and period splits provides the basis for the possible analyses.

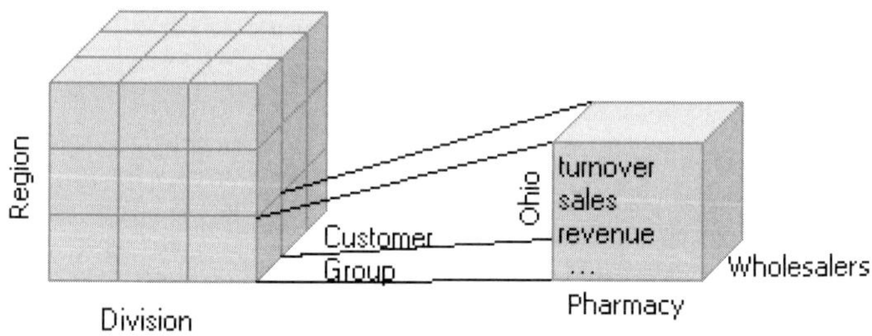

Figure 5-3: Information space as starting point for analyses and planning

The activities from the operative applications (e.g., orders, sales) are used to perform period-dependent updating of the data in the information structures. The definition of the *update* determines the information flow of the data from the operative applications into the information system. A differentiation is made between two types of updating control:

- *General control* by the application using update groups: In the operative application, each update group is assigned to a business transaction or a group of business transactions. An update group is a collection of rules that specify whether and how key figures are to be updated.

- *Fine-tuned control* of updating rules: updating rules are defined for specific key figures. These specify how the data to be updated are linked with the previously updated data in the information structure (e.g., accumulate, replace) and which events in the operative application initiate the updating of these key figures and for which combinations of characteristics this updating is to be made.

In addition to the updating of the data from the operative applications, planning data can be written into the information structures.

The user-defined and the standard supplied information structures form the basis for all analyses in the information system. The *analysis tools* include:

- *Standard analyses* with comprehensive capabilities for list processing and graphical presentation
- *Flexible analyses* with the capability of individual list layout and formula definition
- *Flexible planning* with individually designed planning tableaus, various planning methods, distribution functions, forecasting support and interactive graphics
- An *early-warning system* that supports a decision-oriented selection and testing of weaknesses; it allows searching for exception situations and the early recognition of noteworthy developments
- The *logistics information library* that uses simple search strategies to access key figures within the LIS and which can be used to catalog key figures.

5.2.3 General Analysis Methods

A number of *analysis methods* generally available in LIS, and thus also in RIS, are useful for the preparation of management decisions in retail companies. A fundamental analysis method is *drill-down*: Normally an analysis is started at the aggregated level and refined with detailed analysis at lower levels ("projection"). This allows the degree of the information depth to be varied. Several different criteria can be used for each analysis to provide further details. These criteria correspond to the characteristics of the information structure used as basis for the associated analysis. For example, as part of a promotion evaluation, drill-down can be first used on the "sold quantity" and "sales" key figures to provide detailed information for the "article" characteristic, and then further split into the "store" characteristic and the period (sales date). Because the sequence of the characteristics for drill-down is not defined, very different analysis paths can result. Typical analysis paths can be defined as "standard projection" and then executed "with a mouse click". The characteristic value can be changed in each "projection list" (see Figure 5-4). This supports the navigation in the projection lists and the analysis of dependencies between characteristic values. A change to a view at the same level of detail based on another characteristic can be performed with a projection change. A projection for hierarchies uses an existing hierarchy to perform a detailing. For example, in the promotion evaluation, drill-down can split the "sold quantity" and "sales" key figures along the merchandise category hierarchy down to the individual article.

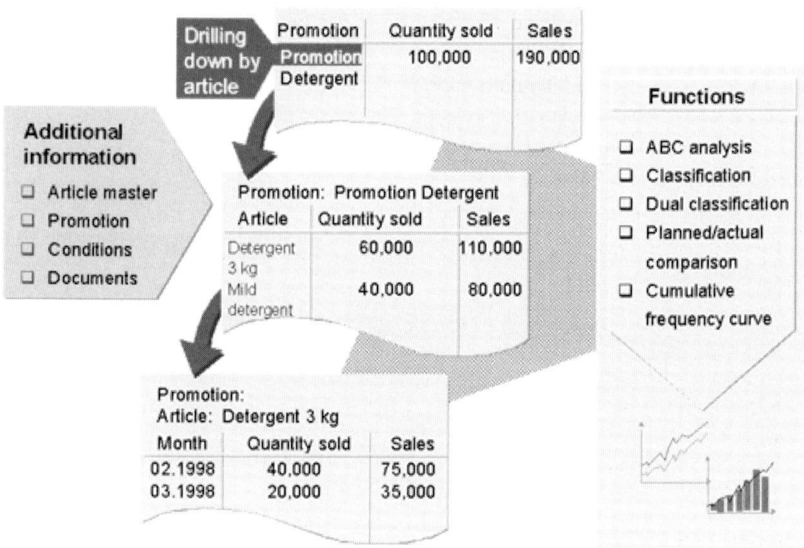

Figure 5-4: Drill-down in the RIS

Other methods that can be used at every level of detail are

- *ABC analysis:* The grouping of characteristic forms with regard to the importance of certain key figures, e.g., assignment of the vendors into A, B and C vendors based on the value of the delivered goods.
- *Classification:* The assignment of the characteristic forms (e.g., the individual stores) into classes based on one key figure (e.g., the area profitability).
- *Segmentation (dual classification):* The assignment of the characteristic forms (e.g., the individual stores) into classes based on two key figures (e.g., area profitability and number of listed products); interrelationships between the key figures and any problem areas become apparent.
- *Top N/Last N:* The ranking of characteristic values (e.g., article) using key figures, for example, the creation of a leader-loser list with the ten best and the ten worst articles based on the sold quantity.

- *Cumulative curve:* The cumulative curve provides information about the concentration of the characteristic values. The key figure values are added in decreasing order. The number of characteristic values and the cumulative values of the selected key figure are marked on the abscissa and the ordinate, respectively. For example, as part of a promotion evaluation, the analysis should be made how the sales for the articles are distributed. An article analysis is performed and the sales key figure selected. Thus, the list shows the articles that were handled together in the promotion and the associated sales key figure. A cumulative curve is produced for this list. An assessment of this graphic could be: five percent of all articles produce 30 percent of the sales. Thus, the cumulative curve provides an overview of the extent to which a large proportion of the total value for a key figure is concentrated on just a few characteristic values.

- *Correlation:* The analysis of the inter-relationships between several key figures. For example, it can be useful to analyze merchandise categories with regard to the correlation between sales and the length of the contact distance.

- *Planned/Actual comparison:* The comparison of the actual values and the planned values for key figures.

- *Representation* of the analysis in list form with the capability to define a user-specific list layout, and to use a range of different presentation graphic types, including time series and portfolio graphic.

5.2.4 Retailing-Specific Standard Analyses

The standard analyses on the basis of standard information structures form the central reporting instrument in the RIS. These can be augmented with flexible analyses using individual information structures. The most important analyses for the retail company are

- *KWER* (short-term retailing profitability statement): This standard analysis supplies data that are of particular interest for sales. It forms the basis for short-term retailing profitability statements. A typical request could be: "What is the situation concerning the value of inventory and the corresponding margins in the stores in a certain region?"

- *Purchasing:* This standard analysis provides data reports that are of particular interest for the purchasing department. A category manager (see Section 7.3) could be interested, for example, in the interrelationships between order volumes and sales for "his" merchandise category for the individual vendors of a purchasing area.

- *Subsequent settlement:* Final settlement and interim settlement results can be compared with each other and so allow the early recognition of any discrepancies between income from the current year and the previous year.

- *Article:* This permits exact article analyses, e.g., analyses of the outflow of goods or creation of leader-loser lists for store and distribution centers. These analyses allow slow-selling articles to be found quickly. The definition of layouts and assigned area schemes permits an analysis of the area profitability. The analysis of article developments over their lifetime can be of interest. Such analyses form the basis for assortment policy decisions to restrict the assortment, to extend the assortment or to change the assortment structure.

- *Promotion:* Analysis of the logistical handling and the selling of promotion goods; selective analysis for individual promotions, types of promotions and themes.

- *Season:* Analysis of seasonal business to correct purchasing decisions for the current season or for the future assortment range. This standard analysis is particularly important for textile retailers.

- *Customer/merchandise category, customer/article:* Analysis of the customer structure at the merchandise group or article level. The customer forms the central analysis object for these two standard analyses. This makes them particularly useful not only for wholesalers, but also for the retail sector, provided that the customer is known.

- *POS statistics:* The "article aggregation / cash register" information structure provides cash register data at the article and condition type aggregation levels. It allows, for example, the analysis which sales for all articles of the "non-alcoholic beverages" merchandise category were achieved on a certain day or what is the sales margin for confectionery for the "Happy Easter" promotion. The "cashier" information structure contains personalized cash register data. Cashier data, such as cash register differences, scanning rates, amount of money in the cash register and cancellation rates, can be evaluated. The data of the "sales per cash register receipt" information structure are used for market-basket analysis, computer based marketing (for known customers) and for shift planning.

- *Market-basket analyses:* These permit the analysis of linked purchases, the general shopper behavior (time-of-day distribution; groups of shoppers, such as bargain hunters, regular customers, customers from competitors; payment behavior, etc.) or the behavior of shoppers for certain assortments (e.g., baby articles) and customer frequency analyses. The cash register receipt is the consumer's "voting paper". Results can be displayed in portfolio graphs, e.g., with "Average sales per buyer in the merchandise category" and "Merchandise-category-specific buyer frequency = number shopping baskets with items from the merchandise group / total number of shopping baskets" as axes. Data mining methods can be used to discover interesting data patterns, e.g., companion purchases, in the cash register receipt data ("70 percent of the buyers of margarine with a high proportion of non-saturated fatty acids also purchase premium fruit juices"). Shopping basket analyses also form a basis for testing the advertising success (Fischer 1997).

- *Open-to-Buy:* This standard analysis is used to evaluate key figures from the limit planning (see Section 5.2.6).
- *Perishables:* This analysis supplies data accurate to the day to monitor the perishable process. For the wholesaling and retail sector, analyses can be performed starting with the purchase order, include the goods receipt and stock transfers from the distribution centers to the stores, through to the sales at the POS.
- *Physical inventory*: This analysis supplies data for the physical inventory controlling. For example, extreme differences prior to booking the physical inventory can be localized using an exception analysis and the appropriate measures, such as recounting, can be initiated.
- *Rough workload preview:* The analyses permit a prediction of the load consideration for distribution centers, provided the SAP LES - Warehouse Management System (see "New Dimension Products" in Section 7.2.5) has been installed. This standard analysis is suitable not only for the retail sector, but also for wholesaler and mail order sector. The analyses can be used, for example, as basis for the shift planning for employees or to optimize deliveries. In addition to an overall summary for all warehouse processes, four further analyses are provided: goods receipt/storage, picking/goods issue, customer/ store returns or vendor returns.

The basis information structure is to be represented using the KWER standard analysis as example. Characteristics that can be aggregated and analyzed are purchasing organization, sales organization, distribution channel, customer region, site type (e.g., store or distribution center), site or merchandise category. A different information structure is used for drill-down to the article level. Key figures are, for example, issue, promotion turnover, promotion turnover margin, stock, inventory adjustment, physical inventory difference, stock margin, order margin, order volumes, annual inventory turnover, range of coverage, returns, sales, sales margin, load transfer, goods issue from the distribution center, vendor goods receipt, distribution center goods receipt or goods receipt margin.

Selection versions are used to maintain and store the data selection for a standard analysis in background processing. The analysis area can be saved as a variant. This allows not only a simpler handling by the end user but also increases the efficiency of the processing and the access. The processing of larger data quantities in background reduces the online operation loading and makes the execution speed for the analysis many times faster than for the online standard analysis. Because selection versions can be assigned to distribution lists, a regular planning of the analysis permits an automated distribution. Optionally, authorizations (using the authorization concept) can be assigned for certain selection versions.

5.2.5 Flexible Analyses

Flexible analyses extend the reporting capabilities of the standard analyses by allowing an individual grouping and aggregation of key figures and characteristics. It is possible to combine several information structures with each other and to calculate new key figures using user-defined formulas. Whereas detailed information can be extracted for the responsible person, aggregated information is provided for management. The specification of a so-called evaluation structure defines the data source of the analysis before the selection of key figures and characteristics produces a customized report. The report layouts can be customized using the report writer. Several layout variants can be used.

Every information structure delivered in the standard already has an evaluation structure with the same name. In addition, an evaluation structure with the same name is generated automatically for all user-defined information structures in the customizing. Any changes to an information structure regenerate the based evaluation structure. If an information structure no longer exists, the associated evaluation structure will be deleted. Consequently, if only one information structure is to be evaluated using the flexible analysis, no evaluation structure needs to be created. Only if data from two or more information structures are to be combined and evaluated a new evaluation structure has to be created with reference to the existing evaluation structures. The analysis then must be defined based on this new evaluation structure.

5.2.6 Planning Support

The comparison of the planning data and the actual data provides an important basis for decision making support in the retail sector. Consequently, the RIS is not only used for the analysis of the actual data, but planning data can also be entered. Various planning concepts can be stored parallel as plan versions. The planning data then can be used as part of the analyses for a planned/actual comparison. The planning is performed not only on the basis of the information structures supplied in the Standard but also on the user-defined information structures. The maintenance of the data is performed using planning tableaus. For simplification, distribution and forecast functions can be used. The following forecast models are available for forecasting: constant model, constant model with customization of the smoothing factor, trend model, season model, trend-season model, moving average and moving weighted average. The view of the elements to be planned can be changed during the planning. For example, the planning data can be entered periodically for several key figures of an article. It is also possible to enter the planning data for one key figure for various articles. User-specific planning tableau layouts, the so-called planning types, can be defined. A planning type specifies the content and structure (layout) of the rows in the planning tableau and the calculation rule associated with these rows. A mouse can be used in an interactive graphic display to change the planning data.

A central element of the planning is the *planning method* used in association with the information structure. Depending on the planning goal in the retail company, the following methods can be used:

1. Consistent planning:

- The data are stored at the level of the characteristic value combination with the highest level of detail.
- Because dependencies exist between the planning levels, changes made in one level have immediate effects on other planning levels. The aggregation and disaggregation of data are performed automatically at runtime.
- The proportions needed for the automatic disaggregation can be determined in two ways, either using previously planned data or through the specification of constant proportions.

2. Stepped planning

- The data are stored at all planning levels.
- The levels can be planned individually.
- The planning levels are independent of each other. Thus, plans at different levels do not need to be consistent.
- It is possible to test and change the data before they are aggregated or disaggregated.

3. Delta planning:

- Data entered at one level can be aggregated automatically at runtime to higher levels.
- Changes made at one level are not disaggregated automatically. The difference between the sum of the individual values at the lower level and the total value at the higher level is retained.

Although analyses can be performed for each required plan version of an information structure, because drill-down would not be appropriate for the stepped or delta planning with their independent planning levels, it can be used only for consistent planning.

An important operational area of the planning system in the retail sector is the *limit planning* OTB (Open-to-Buy). OTB is a scenario that helps purchasers or category managers with the planning and monitoring of their specified purchasing budget. The OTB key figure that shows the remaining purchasing budget as difference between the planned budget and the previously-made expenditures, a critical value here, is calculated at runtime. Sales data, ordering data and goods receipt data are updated regularly. This makes it is possible at any time to see what amount of a budget has already been spent and how much is still available. The budget is divided into the released budget and the reserve budget for subsequent purchases or unexpected changes in the demand. The early-warning system (see Section 5.2.7) that can be used to monitor the continuing use of the purchasing budget automatically generates a warning should there be a danger of the budget being exceeded. The use of consistent planning ensures data consistency. Three successive phases characterize the limit planning:

1. The monthly planning values for sales, closing stock, markdowns and physical inventory differences specified in the *planning phase* form the basis for the calculation of the seasonal budget and the purchasing budget released monthly (released budget).
2. Purchase orders whose delivery date lies in the planning period are entered in the *purchasing phase*. Because all purchase orders reduce the limits, the unused part of the released budget is available at any time.
3. The planning values are compared with the actual values in the *business phase*. The following factors are considered here:
   - good receipts that deviate with regard to time and quantity
   - actual turnover
   - planned-actual deviations for markdowns and physical inventory differences
   - all goods movements to other stores or distribution centers that were not planned and thus not considered in the budget calculation

   This should allow the early avoidance of significant planned-actual deviations.

The business goal of this control is to achieve the planned closing stock for each current period and thus to ensure that the planned-actual deviations do not affect the following periods. If, for example, more sales are achieved in one month than planned, then the closing stock at the end of this month and consequently also the stock at the start of the following month would be less than planned. In this case, appropriate additional purchases must be made to attain the required closing stock. In the opposite case, measures must be adopted to avoid additional stock being accrued as the result of additional purchases; it is also possible to reduce existing stock with a promotion or a goods clearing. The tools for the limit planning have already been mentioned above: the planning tableaus and the "Open-to-Buy" standard analysis. OTB is suitable for business areas with a heavily changing assortment, because these areas are usually planned at an aggregated level rather than at the article level.

A special planning method in the retail sector is used for seasonal goods. It is usual to reduce the price for these articles stepwise during the season (perform so-called "markdowns"). This includes a *markdown planning*, the careful monitoring of the revenues resulting from the markdowns, and a separate handling of the article. As part of the markdowns planning, dates for planned markdowns are specified before the start of the season. To check the success of the markdown strategy, the planned sales are regularly compared with the actual sales during the season. SAP Retail satisfies the following requirements to allow the integration of application systems to support the markdown planning:

- The markdown planning must forward markdown information at specified dates to a central retailing calculation so that the prices for sales goods can be recalculated automatically. The system may need to round prices to specified price points (e.g., $X.99). Markdowns are performed only when they have been activated.
- New prices must then be downloaded into the stores' POS systems using a POS interface (see Section 4.7).
- Finally, to allow the planned and actual sales to be compared, the sales data are forwarded to the markdown planning during the season.

The following points characterize the differentiated functionality of SAP Retail to support the markdown planning:

- The possibility to preplan dates for several markdowns on the basis of percentages of the original selling price.
- The possibility to create standard markdown rules that reflect the calculation procedure. The assignment of these rules to articles allows an optimization of the markdown planning function.
- The possibility to create plans for a range of articles or stores.
- The possibility to assign a standard markdown rule to an article when the planning data are entered and the option to later modify this rule. These Changes do no affect any other articles to which the rule was assigned or will possibly be assigned later.
- An interactive planning tableau can be used to create plans and show their effects of the planned sales. If a markdown percentage value is changed, the planned turnover will be recalculated automatically. Changes can continue to be made until an acceptable result is attained.
- The system warns when overlapping validity periods for markdowns have unintentionally been assigned to an article.

- The actual data are updated automatically. These include not only the sales, but also the range of coverage of the available stock, the current stock quantities and values, the number of outstanding purchase orders, etc.
- The possibility to make changes at any time and also to add or remove price phases (dates when markdowns are to take effect).

Non-SAP systems that can be coupled to SAP Retail also support special planning (refer to Chapter 4 for the general capabilities). For example, StaffWorks from Campbell Software Inc. can be used for the *shift planning*. SAP Retail passes data about the type and extent of the planned deliveries. StaffWorks uses various algorithms for absenteeism, breaks and daily working time to determine the daily shift schedules in the store so that sufficient time to serve the customers remains in addition to the expected loading for handling the goods receipts. Not only can planned physical inventories be taken into account, but also promotions. The aggregated sales and customer data can be compared with the shift planning to adapt the parameters appropriately. In particular, the aspects of the liberalization of store opening hours and seasonal fluctuations in the business procedure are particularly important for this type of planning. We can only mention here, that in conjunction with the SAP-HR personnel administration module, StaffWorks can be used to support other personnel-related tasks in the store, in particular the monitoring of the personnel costs.

5.2.7 Early-Warning

Early-warning systems should provide managers with timely warnings about future risks and chances. In addition to portfolio techniques, scenario techniques and strength-weakness profiles, strategic early warning makes particular use of weakly structured information from external sources, to recognize "weak signals" (Ansoff) in time and deduce handling options. SAP Retail does not support the latter. The range of the New Dimension Products from SAP, a new generation of analytical applications based on the SAP Business Information Warehouse (SAP BW), provides the Business Information Collection (SEM-BIC) component in the Strategic Enterprise Management (SAP SEM) that also permits the integration of unstructured business information from internal and external sources (Meier, et al., 1999). The use of BIC is not restricted to early warning, but has general significance for the integration of external information in MSS (see Section 5.3.4).

Operative early-warning systems are normally based exclusively on heavily structured information. They should recognize exception situations and should signal them in a suitable form to allow the management to recognize any threatening wrong developments and opportunities in time. This permits the management to respond appropriately. As part of the LIS, and thus the RIS, the so-called early-warning system (EWS) supports only the operative early clarification, although effects in the strategic area obviously cannot be excluded. The EWS is based on the previously discussed information structures. The EWS can monitor the information that is updated in the information structures. Exceptions in the form of conditions that apply to key figures and the resulting measures can be defined for each characteristic and every combination of characteristics for an information structure. Whereas exception conditions are always tested at the lowest aggregation level, exception situations accentuate the highest aggregation level. A subsequent processing can pass any recognized exception situations to the responsible employee. Workflow, mail and fax are available for the subsequent processing.

The following conditions can be defined:

- *Threshold values:* A threshold value and an operator (such as >, <, =) based on a characteristic or a characteristic combination (e.g., article/store with sales larger than $ 8000) are defined for a key figure. The choice of the number of forecasting periods also allows the testing of future developments.
- *Trend:* A key figure is tested for the presence of a positive or negative trend (e.g., article/store with a negative trend for sales or a positive trend for turnover rates). A statistical test of the trend is possible when three or more periods are available (a trend analysis is not sensible for fewer periods); so an exception situation occurs only when the trend lies in the five-percent confidence interval of the test.
- *Planned/actual comparison:* Test using the percentage deviation of the actual values from the planned values (for example: for which customers of the wholesaler does the level of meeting the plan for the received orders lie below 80 percent?). With the selection of forecast periods even future developments can be tested.

The EWS can be used interactively in the standard analyses or periodically in background. The concrete testing for exception situations can be done in three separate ways:

- *Standard analysis with exception:* An exception can be specified for performing a standard analysis. A colored marking at the highest aggregation level already indicates the exception situation. Drill-down can be used to navigate to the exception situation.
- *Exception analysis:* When an exception analysis is performed, only the data applicable to the exception are shown. The sequence of the characteristics selected from an information structure determines a subsequent analysis path.
- *Periodic analysis:* The periodic analysis searches through the data files in specific time intervals for exception situations. The occurrence of an exception initiates subsequent processing using mail, fax or workflow. The

periodic analysis can be done in two ways: event or system-driven. For an event-driven analysis, the search for exception situations is performed only when data have changed as result of a logistical event (e.g., goods receipt, purchase order, stock transfer). The system-driven periodic analysis is performed in fixed time intervals (daily, weekly, monthly). The first form is suitable for particularly important characteristics, whereas in all other cases, the test can be performed in fixed time intervals.

5.2.8 Logistics Information Library

The Logistics Information Library (LIL) provides a systematic catalog for all available key figures of the LIS. It permits a fast and direct access to more than 900 key figures from various sources in the standard and an individual grouping of key figures.

5.3 Management Support on the Basis of the SAP Business Information Warehouse

5.3.1 Concept

SAP Business Information Warehouse (SAP BW) can be considered as being a further development of the OIW. Although the information systems of the OIW are components of the R/3 system or the industry solutions, such as SAP Retail, SAP provides with the BW an independent data warehouse solution to extend its product offering as part of the Business Framework. This represents the kernel of the previously mentioned New Dimension Product. The BW is open for a wide-range of information sources. These can be SAP systems, non-SAP systems or external information sources, such as Nielsen, Reuters, Dun & Bradstreet or Internet information. Integrated metadata used with BW and SAP Retail permit a synchronization of the company data and business processes documented in the Retail system with the data available in the BW. These are extracted, transformed, structured, and prepared for analyses and reporting using specific schemes adapted to the associated business processes.

The InfoSources are defined using the InfoObjects, the smallest elements of the information presentation, as basis. An InfoObject can be a characteristic (article, store, vendor) or a key figure (delivery quantity, sales). The InfoSources describe the information input into the BW based on a business-related topic (for example, shopping baskets). The assignment of the InfoObjects to the InfoSources ensures the semantic integrity of the BW. These InfoSources form the basis for updating the InfoCubes. They are multi-dimensional data stores. The InfoCubes are filled with key figures and characteristics from one or more InfoSources, and form the basis for the multidimensional analyses and evaluations (OLAP).

The semantic data model used as basis for the BW is the familiar star scheme (Hahne 1998) with facts table ("key figures") and dimension tables ("characteristic value combinations"). Related characteristics are grouped in a dimension. The dimension tables are linked with keys to the facts table. The time dimension must be used as special dimension (also refer to Figure 5-5).

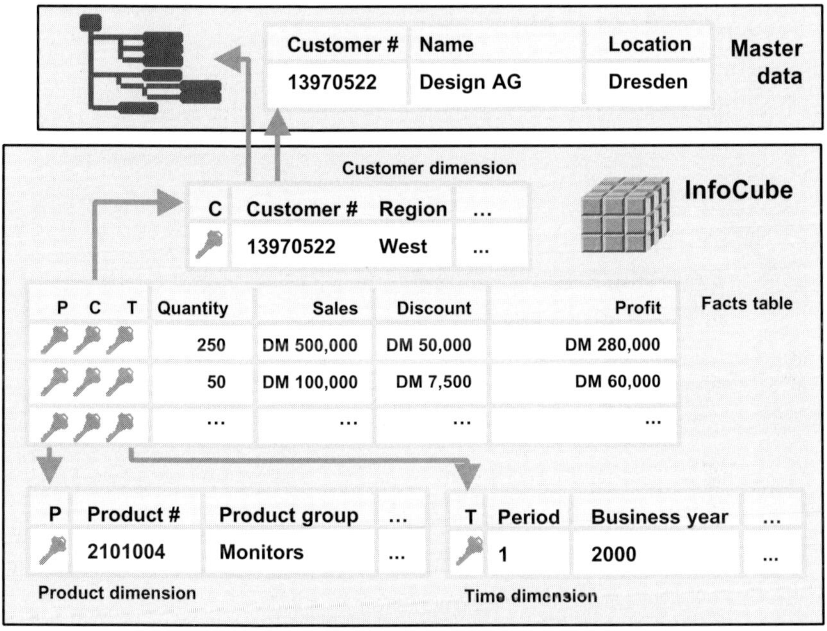

Figure 5-5: Extended star scheme

The Business Explorer is the graphical user interface for the SAP BW; it integrates a report and analysis interface, and comprises the Business Explorer Browser (report management) and the Business Explorer Analyzer (works with the report and builds a report).

The Business Explorer is based on Internet technology and allows the results to be displayed in Microsoft EXCEL. Figure 5-6 shows an example of the user interface. The information is presented to the user as the result of various analyses (queries). A query refers to one or more InfoCubes. Thus, all requested characteristics and key figures must be present in the InfoCubes when a query is created. The storing of the queries in workbooks allows the user to save and manage queries with related content at a single location.

The Business Explorer provides the user with clusters of workbooks with defined standard reports. These clusters also form the basis for a "channel" that can represent a so-called role. A task-oriented role, for example, is that of a key account manager who is provided with specially preconfigured reports in the Explorer. He can use drag-and-drop to add important workbooks to his favorite reports that he personally uses most frequently. A workbook can be assigned to several favorites. A report catalog, including a report preview, makes transparent this information offering. Similar to the familiar Web browsers, the users can surf through the BW and, also with ad-hoc queries, initiate the widest range of analyses. The basis therefore is the OLAP processor integrated in the BW to search and perform calculations in the background. If required, the Business Explorer user can invoke transactions of the operative application system or can access Internet pages.

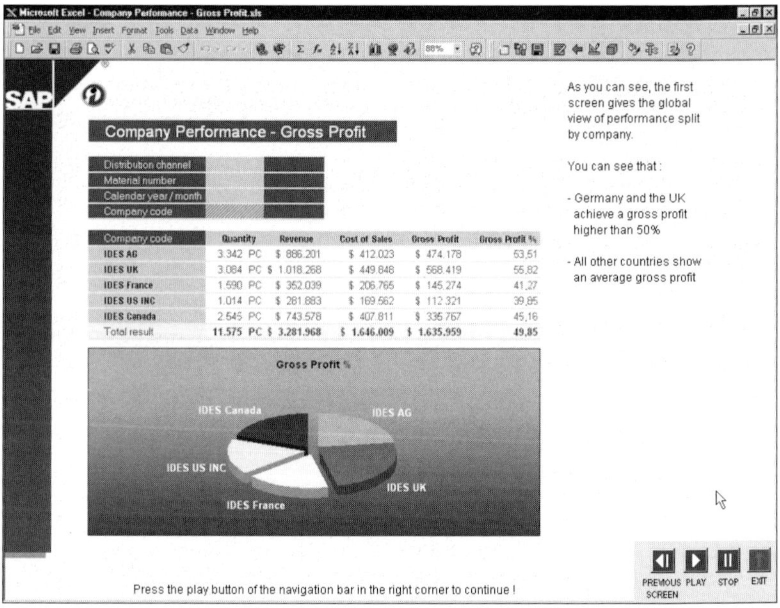

Figure 5-6: Graphical user interface for the Business Explorer

The BW is not an "empty" Data Warehouse, but has already been preconfigured for the important business processes (Business Content, consisting of Standard-InfoSources, Standard-InfoCubes, Standard-Queries and Standard-Workbooks, and R/3 extractor programs). Thus, there exist structures to be filled in the company, which, however, can be previously modified and extended. The data extractor contains appropriately preconfigured data transfer programs. Section 5.3.3 provides an overview of the retail-specific parts of the Business Content. The Administrator Workbench ensures an effective management of the Data Warehouse.

5.3.2 Advantages of the Business Information Warehouse Compared with the RIS

With the completion of the SAP BW 2.0B, SAP recommends the use of the BW instead of RIS/LIS. The current Release 2.1C provides the Retail user with a high-performance environment with preconfigured business models (Business Content). These business models satisfy most business requirements (SAP 2001). Although the use of RIS/LIS remains possible without limitations, its use should be restricted to specific situations and after an appropriate evaluation.

The following reasons favor the use of SAP BW instead of RIS/LIS:

- Whereas SAP BW is designed as standalone product, the RIS/LIS is integrated in the operative Retail System. To ensure an optimum use of the operative system, the data warehouse operation should be decoupled to allow the load on the database server to be distributed over several resources. The operation of the RIS/LIS on a separate system should in future not be favored for reasons of the availability of a SAP Data Warehouse solution, which supports just such an architecture,
- As part of the 2.0B content development, the information content of the RIS is largely represented in the SAP BW. Missing information structures have been added as part of the 2.1C content development. A preconfigured solution has been developed that is compatible to the standard supplied RIS. However, this provides a technological base much more suitable for the analysis of mass data. This solution can be adapted much easier to the customer-specific requirements.
- Significant architectural differences between the technological realizations of SAP BW and RIS/LIS favor the use of SAP BW:
  - Relational tables (information structures) are used in RIS/LIS to save data. However, in particular, the analysis levels needed for the subsequent reporting must be stored redundantly. The data are saved in the SAP BW as star scheme (InfoCube), where the transaction data (facts table) and master data are maintained separately and performance-optimized accesses (dimensions) can be performed.
  - The data model in the SAP BW supports a parallel transmission of master data and transaction data. This means that the performance-intensive concentration of master data, as needed for the RIS/LIS operation, is no longer necessary. In particular, reclassification problems (for example, a change of the article/merchandise category assignment), which require extensive reorganization runs (copy management) in the RIS/LIS, are normally not required when SAP BW is used (exception: building aggregates).

- The data model of the RIS/LIS requires that aggregates are represented in a separate data model (information structure). Without a number of such self-contained "data stores", each of which is subject to a separate supply, maintenance and administration, adequate performance of the LIS/RIS operation cannot be realized. The use of SAP BW allows an unlimited number of aggregates to utilize the existing data model (InfoCube). To analyse the access behavior of the users (queries, drill-downs) system support is provided.
- The standard analysis is the principal means of evaluating data in the RIS/LIS. This provides an inflexible report in which the data read from the database are maintained as an internal table and must be reprocessed/reaggregated for each drill-down. The number of records that can be processed is limited and the user can expect memory/performance problems. In contrast, SAP BW uses OLAP technology that makes automatic use of existing aggregates. This ensures a significantly higher evaluation performance. It is possible to completely transfer the creation of the queries (Ease-of-Use) for BW reports to the user department.
- SAP BW allows data from several operative systems to be collected and analyzed together.

➢ Within the mySAP Business Framework architecture, SAP BW provides a central component which is required for the use of additional mySAP components, such as SAP Strategic Enterprise Management (SEM) or SAP Customer Relationship Management (CRM), whose application development is more or less closely connected with the SAP BW. Thus, the use of SAP BW leaves open the option for the use of these products.

➢ As strategic product of SAP, the SAP BW exhibits significantly shorter innovation cycles than RIS, which no longer will be further developed to any significant extent. This allows a faster response to changed customer needs.

However, these advantages are bought with additional costs that always result from the introduction of an additional software system. This requires an appropriate evaluation for every customer project to determine the extent to which the use of the RIS or the SAP BW is sensible. It is also possible to envisage a phased introduction in which the RIS is used initially for operative management support and only in a subsequent second phase SAP BW is implemented to provide an increased strategic operative management. To take load from the operative system, in this case the majority of the RIS tasks can be represented in SAP BW. In any case, the use of the SAP BW is the more expandable solution for larger companies in the retail sector.

5.3.3 Retail-Specific Business Content

Release 2.0B extended the Business Content supplied with the SAP BW and regrouped it according to roles. The preconfigured Business Content supplied for the Retail industry solution for the first time provides a large part of the RIS functions in the BW. The content has been grouped into the InfoArea Retail. The following section discusses the provided specific preconfigured functions.

The Business Content supplied with SAP BW should be considered as being a suggestion from SAP as part of a best-practice scenario. It is not possible to provide ready-to-go content in such a heterogeneous industry like retailing (retail sector / wholesale / mail order, foodstuffs / fashion, etc.) with greatly varying information requirements of the individual retail companies. In contrast, the supplied data extractors, which form the key for a fast and economical implementation of SAP BW, do not depend on the company and so can be used without needing to be modified. However, the attempt was made to cover the main information requirements of all retail sectors and thus significantly simplify the customer-specific content development.

Unless otherwise indicated, all the following statements apply to SAP BW Release 2.1C.

Data extractors

The data extractors represent a ready-to-go solution to transfer the operative data from the SAP Retail System to the InfoCubes of the Data Warehouse. They consist of two parts, PlugIn and InfoSource. The PlugIn is an add-on development that can be integrated without problem in the existing Retail System. The data won using the PlugIn is made available to SAP BW by InfoSources. Support is provided for the extraction of master data, transaction data and metadata.

The extractors needed to obtain the movement data are based on the logistics information system. However, here in contrast to the RIS information structures, only the data records not previously transferred to the SAP BW are stored. This removes a significant load from the source systems. The following data are analyzed for the retail-specific Business Content:

- purchasing data
- stock data (only for the initialization)
- inventory management data
- sales data (POS cash register receipt and cashierdata)
- revaluation data
- planning data from assortment, store and merchandise planning

Other movement data extractors in customer projects can be obtained easily from the existing LIS/RIS information structures.

Retail-specific master data are obtained from the following application components:

- general logistics (material data, site data, assignment site/material)
- general materials management (vendor)
- sales (customer, billing data)
- control (actual costs of cost centers)
- personnel management (settlement data, target working hours, actual working hours)
- retail industry solution (promotion, assignment promotion/articles)

InfoCubes

The Retail InfoArea has been subdivided into the subareas: merchandise and assortment planning, retailing, point of sale and store controlling, which contain predefined InfoCubes. Although these are already fully functional, they can be also used as basis to produce customized InfoCubes.

➢ In the subarea merchandise and assortment planning diverse InfoCubes for a best-practice scenario are predefined. They serve as a reference for own merchandise and assortment planning scenarios and are not to be presented in detail in this context.

➢ The InfoCubes provided in the retailing subarea cover the typical information requirement for the user of the retail information system. Most of the important key figures from the retailing cycle are defined in these InfoCubes and can be evaluated as queries.

  - The article InfoCube represents comprehensively the most important business content of the various the retailing areas. This should allow a comparison of key figures from the purchasing, inventory management and sales. In particular, the stock key figure is modeled in this InfoCube.
  - The InfoCube goods procurement has been created to represent the specific business content of the contracting. It focuses on the vendor-related controlling (on the basis of purchase order, goods receipt, invoice).
  - The customer and consumer InfoCubes have been created to represent the sales business area. Whereas the first InfoCube has been designed for the wholesale sector and all customers are considered as a whole, the second InfoCube concentrates on the consumers. This orientation makes the InfoCube suitable for mail-order companies and retailers with many

known customers. The different target groups of the two InfoCubes have been taken into consideration for both the modeling of key figures and characteristics, and for the choice of the navigation attributes (e.g., only the consumer InfoCube has age and sex as navigation attributes).

- The goods flow InfoCube represents the business content with regard to the goods flow control in goods distribution centers. This allows the effects of the various processes for the passage of the goods through the distribution center (cross-docking, flow-through, warehouse removals) to be analyzed with regard to the logistical costs and the availability of the goods.
- All stock changing goods movement in the logistical procedure, such as goods receipt and goods issue, inventory adjustments and returns are reflected in the InfoCube article movement.
- Stock data can be traced according to the desired degree of detail accurate to the week or month in the InfoCube stock. In this context also the purchase price revaluation and the revaluation at retail of article stocks on value basis are considered additional to article receipt and issue.
- The InfoCubes purchase price revaluation and revaluation at retail represent these aspects of inventory management in retailing.

➢ Five InfoCubes have been created for the point of sale subarea. These represent an initial extension for the content development in the POS area.

- The cash register receipt data (atomic) InfoCube represents in the SAP BW the familiar "cash register receipt related sales" information structure from the RIS. Because this InfoCube contains the atomic POS base information (store, cash register receipt number, article, EAN, conditions, payment), it forms the basis for a wide range of analyses. Two additional InfoCubes (cash register receipt and cash register receipt/article) are provided for further shopping basket analyses.
- The cashier InfoCube represents in the SAP BW the "cashier" information structure also known from the RIS. With the additional consideration of cash register receipt data, this InfoCube serves as basis for a qualified sales audit that can also be performed over a longer reporting period.
- The InfoCube point of sale control enables comparisons in the sense of a sales audit, which unites the atomic cash register receipt data with the cashier data and exposes variances between the collected cash register receipt values and POS accounting values.

➢ In the subarea store controlling all the necessary information from logistics, financial accounting, controlling and personnel managment are focused for the purpose of a comprehensive store controlling. This subarea contains the following InfoCubes:

- Costs which arise in stores or are accountable to them are registered in the InfoCube store costs.
- Logistical key figures, which arise in a store are provided in the InfoCube logistics.
- The necessary data for the comparison of planned target working hours with actual working hours are provided by the InfoCube personnel hours. The analysis of settlement data accurate to each employee is provided by the InfoCube payroll.
- All these data and the cash register receipt data are combined and can be compared with each other in the InfoCube store controlling.

More advanced InfoCubes can be created to meet customer requirements using the retail-specific and industry-neutral extractors.

Roles

Considering the goal of needing to satisfy the specific information requirement for different persons, the orientation of the contents to the characteristic roles in the retail companies is a central aspect for the development of MSS on the basis of the SAP BW.

The roles are defined at the smallest granularity. Thus, it can be expected that in small companies a single person may need to assume several roles. In this case, a multiple assignment should be performed during customer projects.

A role is the grouping of all tasks that should be performed by a human operative as a whole. Although a job can comprise several roles, the work involved in a role cannot be divided.

To be able to perform the tasks of a role, the specific information requirements of the human operative must be satisfied. This requires reports adapted to the role.

These roles have been realized for SAP Retail within mySAP Workplace:

- customer advisor
- retail category manager
- department manager
- distribution specialist
- good receipt specialist
- retail purchasing manager
- retail sales manager
- retail stocks planner

- store controller
- store manager
- transport manager
- warehouse manager
- wholesale sales manager

However, many of the provided queries can also be used for additional customized roles.

It can be expected that the importance of the role in the product management of SAP will tend to increase with the consequence that the future licensing may be made cross-product based on roles.

5.3.4 Integration of External Information

Although the integration of data from external sources, such as online databases or the Internet, is not trivial, retail companies live from and for the market. Consequently, the most important information for the company management comes from outside. Key figures from company-internal sources often attain their full importance only with reference to the company environment (Mertens 1999). One of the leading worldwide providers of external business data that cooperates with SAP is Dun & Bradstreet Corporation (D&B). The "D&B for SAP R/3" product provides European SAP customers with online access to business information from approximately 17 million European companies. Retail companies can use the data, for example, to optimize the purchasing behavior. For example, so-called hierarchical company relationships (family trees) can be recognized and used to realize better purchasing conditions which can result in the realization of previously unused saving potentials. Combined with SAP BW, this provides interesting new analysis possibilities.

A further cooperation with ACNielsen promises to be particularly interesting for the consumer goods industry and retailing. In future, it should be possible to couple the BW with three ACNielsen services: ACNielsen SCANTRACK (contains scanner data from the retail sector), ACNielsen Homescan (a consumer panel based on weekly data from 52,000 households), and ACNielsen Workstation Info-Server as platform for access to all information offered by ACNielsen. These three services provide the BW user with a form of virtual data mart for the integration of marketing information from ACNielsen with the SAP BW while providing the complete functionality of the BW. Colgate-Palmolive Co. in New York was the first pilot user in the second quarter of 1999.

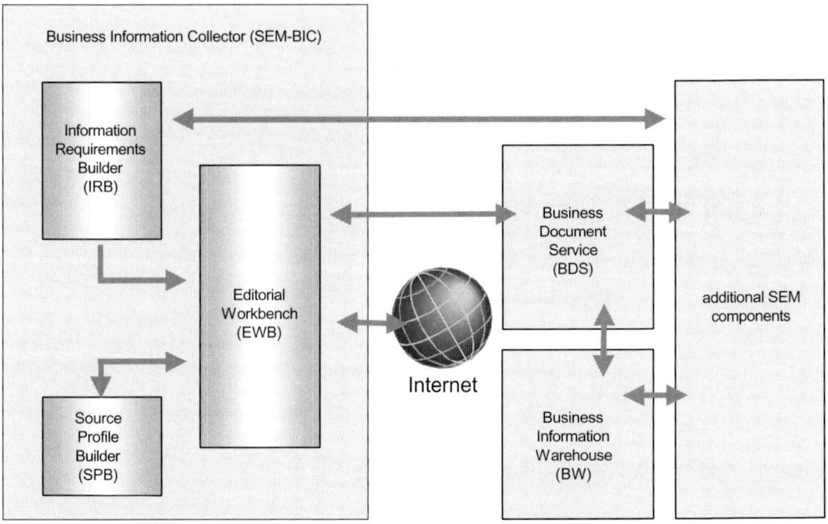

Figure 5-7: Architecture of the Business Information Collector (based on Meier et al. 1999)

The BIC (Business Information Collector) component already mentioned in Section 5.2.7 provides particular support for the targeted search for management-relevant information in the Internet, the filtering, the matching and the combination of the search results with heavily structured internal data. Figure 5-7 shows the basic architecture of the BIC. The Information Requirements Builder (IRB) and the Source Profile Builder (SPB) supply the Editorial Workbench (EWB) with information requirements as search requests, and metadata for source selection and query generation, respectively. The EWB supports both the search in the Internet, and the filtering and linking of the search results with internal data. It is possible to transfer newly discovered data sources from the EWB into the SPB. The Business Document Service (BDS) can be used to manage the found information units. Metadata from the SAP BW are also used to index quality information units. This permits a logical linking of unstructured or weakly structured information, for example, from press announcements about a competitor, with highly structured data, for example, from market research.

6 Electronic Commerce in the Retail Sector

6.1 Introduction and Explanation of Terms

Scarcely a development has influenced the structures of our economy so lasting and in such a short time as the new capabilities of the digital business handling over worldwide networks that are covered by the glittering term *electronic commerce*. Even when the predictions about the size of the expected online sales differ, they all forecast for the next few years a significant increase of the sales revenues that will be achieved over the Internet. The largest potential is seen in the electronic business conducted between companies (B-to-B), with a forecasted sales volume of 153 billion dollars in 2000. A significantly lower revenue potential is seen in the online sales to end customers (B-to-C). Figure 6-1 compares forecasts from various sources.

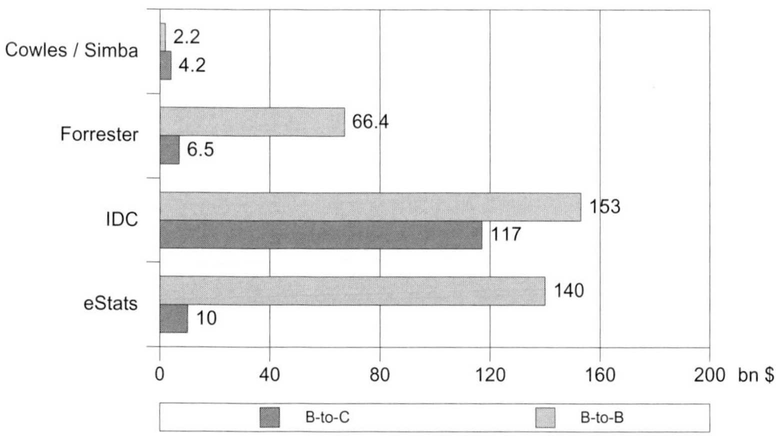

Figure 6-1: Expected online sales in 2000

Depending on the viewpoint and the considered object, the literature provides quite different interpretations of the term electronic commerce (EC). Whereas some authors understand it to be just the electronic transfer of business data, the EC term in the nowadays dominating interpretation serves more as frame that covers the comprehensive support of all value chains for companies through the use of telematic systems. EC is considered as being "... *a concept for the use of specific information and communications technologies for the electronic integration and interconnection of different value chains or cross-company business processes"* (Wolf, et al. 1997). This interpretation will be followed in the following sections.

The adaptations of the organizational structures and business processes associated with EC will significantly influence the retailing sector in the future. However, in contrast to the consequences of the previous developments in information and communications technology, here, not only the relationships between the companies will be changed, such was the case for the use of Electronic Data Interchange (EDI). Rather, EC will influence not only the structure of the relationships of the actors along the value chain but also the relationships between each of the participants of this chain and external partners, and the competitors (see Figure 6-2).

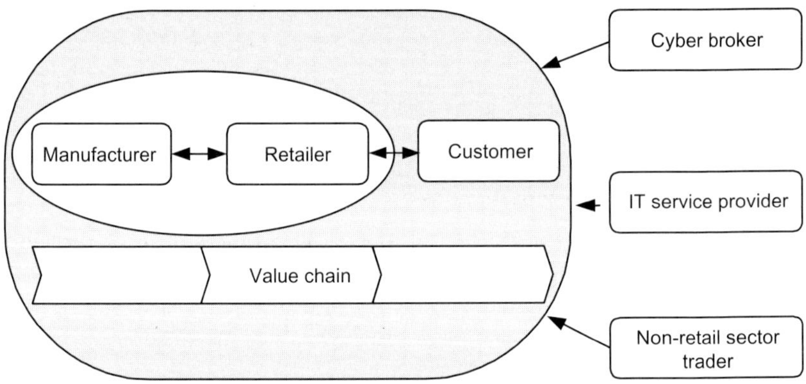

Figure 6-2: Participants and relationships of a market system

The DP landscape previously characterized by proprietary systems often demanded high investment and maintenance costs from companies, which hindered, in particular the integration of smaller companies in the electronic procurement and sales structures of wholesalers or manufacturers. Heterogeneous systems resulted in numerous stand-alone solutions with media fragmentations in the information transfer and longer processing times producing increased costs. In contrast, the establishment of the TCP/IP protocol family as standard for the information transmission and the ubiquity of the Internet provides a number of advantages. Namely, it simplifies, increases the flexibility and the economics of the integration of IT systems and the automation, and the complete DP support for market transactions between the retail companies and their suppliers.

Completely new potentials for handling and extending the market in the end-customer area arise when the technical possibilities of the Internet and of the resulting World Wide Web (WWW) are exhausted through the use of innovative business models. By integrating the point of information, point of sale and point of service, a company's Web site can be used to provide a customer with an individual and theoretically unlimited product offering, 24 hours a day, seven days a week, using a simple-to-use interface. The company Web site can be supplemented with additional services (discussion forums, events service, complaints service, financial services, etc.). This gives the customer significantly more value than that provided at a fixed selling location.

However, the expected closer customer binding between retailer and consumer through the provision of appropriate services is being continually threatened in the Internet as result of the low barriers to market entry, the unimportance of the geographic separation between the party offering services and the party requesting services, and through the presence of (new) competitors. Worrisome for a retailer are also scenarios in which the marketing and distribution of goods and services can in future be increasingly realized by retailing companies, even newcomers to the industry, such as IT companies or specialized cyber-brokers. EC provides the new competitor companies with the capability to substitute not only individual elements but the complete traditional retailing value chain. It can be too little for established retailers only to react to these changes; what is needed is a far-sighted, strategically managed action. Together with reorganization measures, modern standard software offers a suitable platform here.

After introducing the basic technological aspects in Section 6.2, Section 6.3 discusses questions of the market organization. The representation of business aspects in Section 6.4 is augmented with the presentation of the capabilities of the realization of EC concepts using SAP software in Sections 6.5 and 6.6.

6.2 Technological Aspects

After considering a number of standards, which, under the generic term, Electronic Data Interchange, have been significant for the use of information technology for business-related data exchange, this section presents a few fundamental Internet concepts. Finally, a selection of options to ensure the security of the data exchange and the DP systems against external attacks will be discussed.

6.2.1 Electronic Data Interchange (EDI)

A large part of the business documents, such as purchase orders, forecast delivery schedules, delivery notes, invoices, payment transfers, etc., are nowadays created electronically. It does not appear to be very sensible to print, copy and send these documents, processes that are not only much too time- and cost-intensive, but also prone to errors, and then have the data re-entered by the business partners. *EDI* designates procedures for paperless business traffic. A common definition for EDI is: "... the transmission of structured data using specified message standards from one computer application to another using electronic means and with a minimum of human intervention" (CCG 1999).

Up to the mid-1980's, the business documents were exchanged electronically using proprietary data exchange formats and communications standards agreed bilaterally between individual businesses. However, the increasing international competition, as well as new production and logistical concepts in the following years, required a better integration of the vendors in the production processes of the manufacturers and the distribution processes of the retailers. This required the definition of many national and international industry-specific document exchange formats (see Table 6-1). These agreements are based on standardized message types defined for the various forms of the previously mentioned business documents. EDI converters that can be coupled as EDI subsystems using interfaces to standard software, such as SAP Retail (see Chapter 4), are used to convert the data from or into the in-house systems of the business partners. Standard formats are used for the communication.

Particular important for the retail sector in Germany was the *SEDAS* Standard (standard regulations for the standardized data exchange systems) that was published in 1977 by the Centrale for Coorganisation, the joint rationalization association from retailing and industry established in 1974, and which was further developed in the following years. This Standard retains its importance even today.

Table 6-1: Selected industry-specific exchange formats (Deutsch 1995, p. 31)

Exchange format	Industry
DATANORM	Building materials
ELDANORM	Electrical industry
ODETTE	Automobile industry
RINET	Reinsurance
SEDAS	Consumer goods business
SWIFT	Banking
VDA	Automobile industry

The remaining incompatibilities between the exchange formats of the various industry sectors resulted later in the definition of the ISO Standard *EDIFACT* (Electronic Data Interchange for Administration Commerce and Transport). The development of many subsets catered for the requirements of the individual industry sectors (see Table 6-2). For system economics reasons, only those elements of the EDIFACT Standard actually needed were implemented.

Table 6-2: Industry-specific subsets of EDIFACT (Deutsch 1995, p. 47)

EDIFACT Subsets	Industry, Business Segment
CEFIC	Chemical industry
EANCOM	Consumer goods business
EDIBDB	Building materials
EDICER	Ceramic industry
EDIFICE	Electrical industry
EDIPAP	Paper industry
EDITEX	Textile industry

The *EANCOM* Standard is particularly relevant for the retail sector. EANCOM is an EDIFACT subset based on the EAN identifications: In EANCOM messages, every product is identified with a unique EAN standard article number (EAN) and every partner with a unique international location number (ILN). This connects the scanning and EDI base technologies. The CCG has developed conversion and customization aids that allow the conversion from the established SEDAS format into the EANCOM format (and back). EANCOM '97 currently provides 42 standardized message types to exchange business data (CCG 1999).

EDIFACT and its subsets document the purpose of the data exchange with the names of the corresponding message types: for example, QUOTES represents the acronym for a quote, PRICAT represents a price list, and INVOIC represents an invoice. To reduce the transmission costs, messages to the same recipient can be combined to form message groups and then sent as a single transfer file.

Figure 6-3 shows how the files can be transferred to the recipient using either a point-to-point connection (I) or buffered using the VAN (Value Added Network) of one (II) or more (III) service providers, such as IBM or Harbinger. Depending on the connection type, the FTAM (File Transfer Access and Management) or the X.400 Message Handling Service ISO services are used for the data transmission.

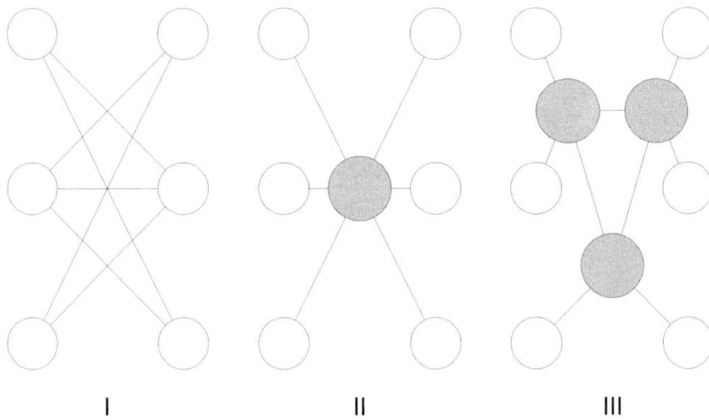

Figure 6-3: Data transmission alternatives in EDI networks

However, the development complexity of the different message types and the high implementation costs for small and mid-sized companies have meant that EDI/EDIFACT has not achieved the decisive market penetration. Nevertheless, the wide use of this Standard within large companies and the requirement of investment protection ensures a continuing support of the associated systems and their inclusion in the Internet applications of the firms. Various concepts are available for the integration of EDI and Internet systems with Web-EDI applications, for example:

- using the encapsulation and identification of the EDI messages as MIME-conform data types (Shih et al. 1997),
- by replacing the EDI segment identifiers with control information (tokens, tags) of semantic document descriptive languages, such as XML (eXtensible Markup Language) (Bryan 1997), or
- by Java applications that interpret the segment identifiers and can display the data correctly (Buxmann et al. 1999).

6.2.2 Internet Fundamentals

The Internet encompasses individual computers and individual networks throughout the world that are connected and communicate with each other on the basis of the TCP/IP protocol family. TCP/IP is connection-free and packet oriented. This means that the sender divides the data stream to be sent into individual data packets that the recipient then reconstructs. To ensure the correct reassembling, each data packet contains details about the sender, the recipient and data fields that uniquely identify the sequence of the packets. An internal counter terminates any data packets that cannot be delivered.

A unique so-called IP-number (e.g., 191.10.12.114) is used to identify and address the individual computers in the Internet. To simplify the handling of the IP-number by humans, it is possible to assign a domain name to a number. The domain name consists of:

- the name of the top level domain that represents a country code (such as .de or .ch) or a usage-related combination of alphabetic characters (such as .com, .org, .edu or .mil),
- further subdomain names (such as spicetrading), and
- possibly a user ID.

For example, the Internet address jk@purchdep.spacetrading.de describes a hypothetical e-mail address of the employee John Klein (jk) who works in the purchasing department (purchdep) of the company Spicetrading. The computer that receives the e-mails is located in Germany (de). However, to simplify the handling of e-mail addresses, increasingly only a subdomain is specified in the e-mail identification. The e-mail address then could be: john.klein@spicetrading.de.

The Internet protocols of importance for the user include:

- the *File Transfer Protocol* (FTP) that permits the transfer of files between different computers
- the *Simple Mail Transfer Protocol* (SMTP) that standardizes the transmission of electronic messages (e-mails) in a TCP/IP network
- *Telnet* that allows you to use a terminal emulation to connect from a personal computer to a mainframe computer
- The *HyperText Transfer Protocol* (HTTP) that uses the Uniform Resource Locator (URL) to permit the addressing of an arbitrary (hypertext) document in the WWW. Together with the HyperText Markup Language (HTML), it was decisive for increasing the acceptance and the commercialization of the Internet.

6.2.3 Security Concepts

The data exchange over the Internet and the connection of a company intranet to the Internet represent a significant security risk for a company. An Internet connection gives not only company employees access to data outside the company but also allows external persons access to the transmission data and to internal information not destined for the public. Thus, appropriate security concepts should be used to ensure not only the confidentiality, the integrity and the availability of the data, but also the incontestability of performed transactions.

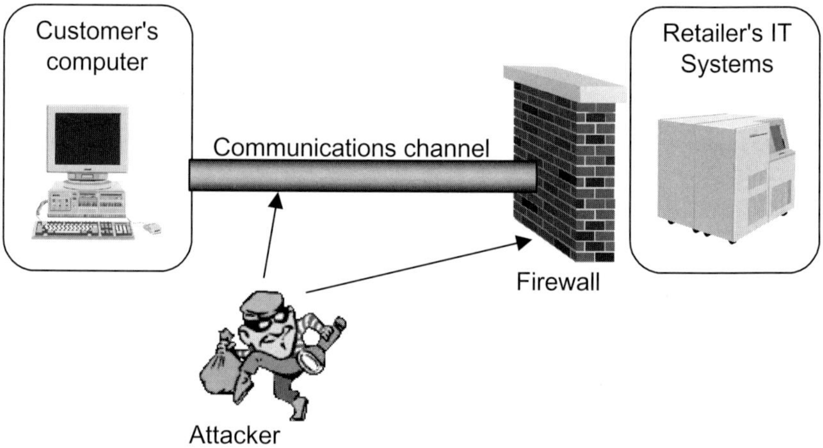

Figure 6-4: Forms of attack on the IT infrastructure

The solution concepts developed here can be divided into two groups, concepts to secure the communication channel (see Figure 6-4) and solutions to secure the company-internal IT systems against unauthorized external and internal access attempts.

Secure Sockets Layer (SSL) is the most popular protocol to secure the transmission path between the external computer, such as for a customer, and an EC system of a retailer, and is used, for example, by many online providers for the secure transmission of credit card details. Both Microsoft Explorer and Netscape Communicator support the protocol. There are also implementations that allow existing software applications to be extended with the functionality provided by SSL (Nusser 1998, p. 125). The functionality of the SSL protocol includes:

- authentication of server and client
- guarantee of the confidentiality of data through their encryption (key lengths between 40 and 1024 bits can be used)
- guarantee of the data integrity
- support of compression methods prior to the encryption.

Secure-HTTP (S-HTTP), another standard to protect the exchange channel, which, in contrast with SSL, pays more attention to ensuring the incontestability of the transaction participation. S/MIME (Secure/Multipurpose Internet Mail Extensions) can be used to secure the transmission of e-mails.

The applications that protect the company network from external attacks known under the term *firewall* can be divided into filtering and proxy firewalls depending on the method used to validate the data exchange. Filtering firewalls are normally realized by a router. They validate the data packets and services using predefined rules. In contrast, proxy firewalls are implemented as a software application that acts as intermediary between the internal and external computers to validate the passed protocol commands. Because both concepts exhibit different advantages and disadvantages, in practice they are normally combined to provide an integrative security architecture model (Chapman et al., 1995, pp. 57).

The creation of a comprehensive security concept requires exceptional care and detailed special knowledge. Even the smallest lapse can cause gaps in the security. For example, for a while any experienced Internet user could execute a small script to gain access the contents of the customer contracts (name, address, telephone number, bank account, size of the household and annual electricity consumption) at the German energy provider Yello (www.yellostrom.de) (Heise 1999). Another example is the Amazon subsidiary Bibliofind.com, where data provided by 98.000 customers have been affected by hacking (Abreo 2001).

However, because the trust of the customers in the security of the IT infrastructure can be critical for the success of an EC presence, the implementation of an appropriate solution should be given only to a company with proven competence in this area.

6.3 Organizational Aspects

The discontinuation of traditional distribution chains as predicted by some authors, and feared by the retail sector, has not eventuated to the forecast extent. However, it can be observed that firms which originally considered their Internet presence only as a means of company presentation, increasingly see this also as an option to establish a direct distribution channel. In addition, traditional retailing companies are being threatened by new market participants or companies that vertically expand their value chain. This section uses the representation of possible participation models of the EC to discuss potential forms of the disintermediation and reintermediation of the retailing value chain.

6.3.1 Participation Models of the EC

Participant groups on classic and electronic markets are public institutions, companies and end consumers. Even when each of these groups can theoretically assume a role as provider, consumer or broker of a market service, the following discussions concentrate on the position of the broker performed exclusively by a (retailing) company. Starting with the interpretation of the electronic commerce as a concept that provides information and communications systems to support all company processes along the value chain, the following section discusses not only bilateral business models (business-to-business or business-to-consumer), but also trilateral business models, such as business-to-business-to-business or business-to-business-to-consumer. Figure 6-5 shows the resulting participant schema. The usual abbreviations are used; administration (A), business (B) and consumer (C).

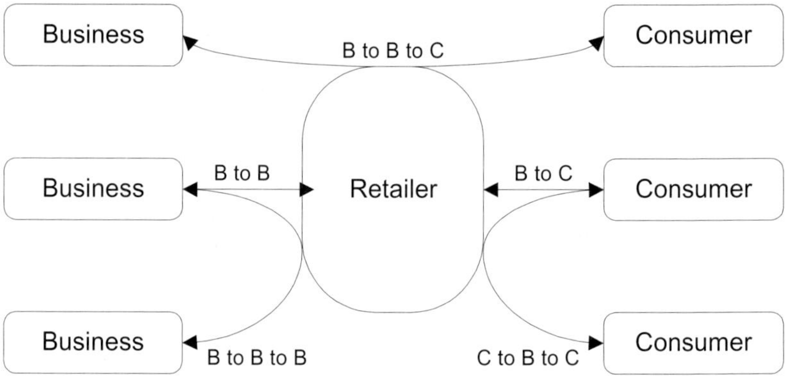

Figure 6-5: Participant model of the Electronic Commerce

Business-to-consumer (B-to-C)

When a manufacturer establishes an eShop application, it has the opportunity model to by-pass the retailer and to sell its products directly to the private consumer. One of the best-known companies that follows this concept and sells its products very successfully is Dell Computer, which realized nearly 50 percent of its $25.3 billion revenue in 2000 using the Internet as a direct distribution channel.

Manufacturers of products that can be described and characterized very easy (e.g., computers) are more likely to follow this concept. On the other hand, manufacturers of goods that require an extensive advisory service or more support time still try to avoid eliminating retailers from the value/distribution chain. Their part as product aggregator or client advisor is still necessary – even in the digital economy – and will therefore be described in the next section.

Business-to-business-to-consumer (B-to-B-to-C)

Whereas stationary retailing is characterized by having a limited sales area and range, high costs for tied-up capital and personnel, together with increasing rent and leasing costs, an Internet shop, as result of the missing physical presence and the high degree of automation for the sales handling, offers significant cost-reduction potential. The option of being able to assign every sold product to a buyer gives the retailer new possibilities for the market processing and customer segmentation.

The B-to-B-to-C model is the equivalent to the retail model for traditional markets in which the retailing company assumes the central role as intermediary between the manufacturers and the consumers. Here the business action for both purchasing and selling is handled over TCP/IP-supported communications channels.

The business model of Amazon (www.amazon.com) can be considered as being a successful example of a company that can be assigned to this participant model. The company founded in 1995 has distinguished itself as being the Internet's leading retailer. In the meanwhile, it lists more than 28 million unique items in categories, such as electronics, kitchen products, books, music, DVDs, videos, camera and photo items, toys, software, computer and video games, tools, hardware, lawn and patio items, and serves more than 25 million customers throughout the world and forecasts sales between $950 million and $1.05 billion for the year 2000. It's revenue expectations for 2001 are approximately $4 billion.

Whereas the sales for online stores show significant growth rates, the attempt to establish virtual shopping malls has not yet been very promising. For example, because of low acceptance, IBM had to close already in July 1997 it's World Avenue opened in August 1996. The shop offering provided by the German shopping24-Mall with only 23 providers can also be considered unsatisfactory in comparison with that of a real shopping center (www.shopping24.de).

Other firms that offer products to their customers over an EC system, are, for example, the retailers Karstadt (www.my-world.de), Kaiser's (www.kaisers.de) and Wal-Mart (www.wal-mart.com), and the mail-order firms Quelle (www.quelle.de), Otto (www.otto.de), and Sears, Roebuck and Co. (www.sears.com).

Business-to-business (B-to-B)

In the participant schema shown in Figure 6-5, the B-to-B-model covers all transactions in which the retailing company itself makes use of services. The potentials of EC in this model concern primarily the optimization of the procurement processes in which the retailer itself appears as an end consumer for goods or services.

Suitable products for an electronic procurement handling include goods that have a low price but a high ordering frequency, and goods that are easy to describe and classify (e.g., office supplies). Services that a retailing company can purchase or contract via an online system include:

- translation or documentation services
- market research services (market, customer or competitor analyses)
- IT services (e.g., software purchase or update).

Sections 6.4.1 and 6.6.2 provide a more detailed discussion of electronic procurement.

Business-to-business-to-business (B-to-B-to-B)

The functionality provided by EC systems makes it possible for a specialized retailer to decisively extend its product offering as intermediary between business customers. A key concept here is the establishing of electronic networks or electronic markets that allow retailers to augment the existing physical value chain with an additional virtual/information value chain. Examples of possible services include:

- procurement services (e.g., by checking the "procurement" catalogs of potential vendors or the tendering pages of public institutions)
- consulting services (e.g., through the provision of special know-how providers when necessary)
- IT services (e.g., through the provision of customized groupware applications to support the negotiations between providers and customers)
- financial services (e.g., through the aggregation of the financial flows and the appearance as financial intermediary in the name of the involved firms)
- marketing services (e.g., by adopting a role as central customer contract contact partner for market investigations)
- sales services (e.g., grouping various product offerings in an electronic product catalog).

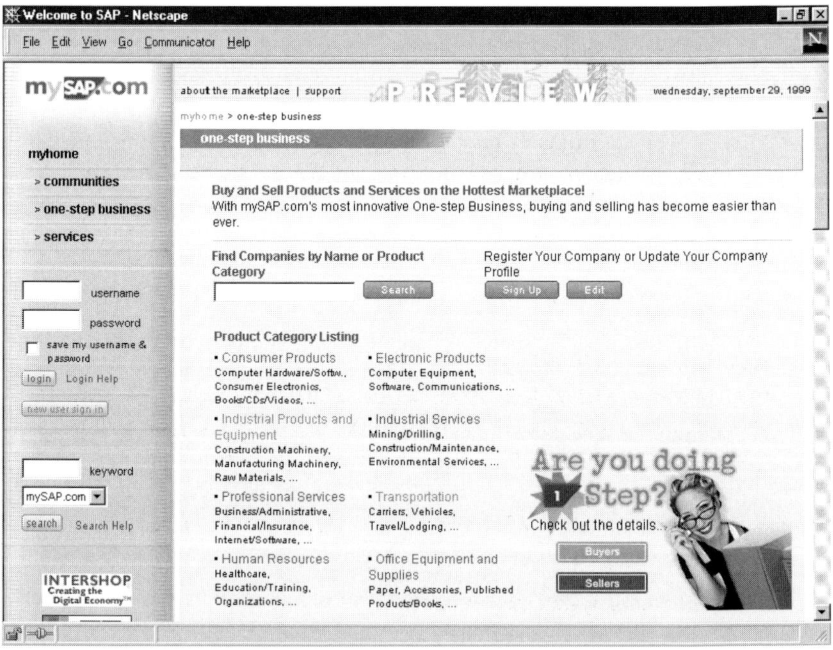

Figure 6-6: Example of an electronic B-to-B-to-B market

Concrete examples of electronic marketplaces for companies are "corProcure" (www.corprocure.com) (see Figure 6-6) or the "MarketSite" (www.marketsite.net) procurement platform established by SAPMarkets and CommerceOne, respectively.

Consumer-to-business-to-consumer (C-to-B-to-C)

In this model, the intermediary adopts an aggregating and coordinating function between the end-consumers. Although the economic importance of this scenario is not immediately obvious, it is of fundamental importance for business models of the EC financed by advertising. The takeover of the Geocities online communication platform by Yahoo in the February 1999 for the purchase price of $4.9 billion is an example of this relevance.

Whereas the first C-to-B-to-C systems were still used as pure communication platforms based on the concept of the virtual communities from Armstrong and Hagel (e.g., Geocities or The Well), an increasing commercialization of the contents can also be observed in this area (cf. Armstrong, et al. 1995, pp. 128). Data mining can be used to extract data from such communities that then can be aggregated to form user profiles. These user profiles are of particular interest for the operators of such platforms and their advertising customers to provide the basis for a one-to-one marketing. For example, the efforts of an increasing number of companies to establish an end-customer-oriented platform underscore the business potential expected from this participant model. The offerings of the ProSieben television broadcaster (www.redseven.de) and FortuneCity (www.fortunecity.de) can be mentioned as examples. The increasing competition and the growing market power of the end-consumer makes it probable that the importance of this participant model will further increase. These participant models also include auction platforms, such as eBay (www.ebay.com) and Ricardo (www.ricardo.de).

Business-to-employee

Business-to-employee (B-to-E) can be considered as being a special form of the B-to-B-participant model. B-to-E applications can be classified as being an option to reduce personnel costs and increase the personnel efficiency that allows employees themselves to manage personal-related business activities or details. These can include:

- change of address details after moving the place of residence
- display a change of the tax class and the family status
- booking of company-internal seminars.

Business-to-administration

Business-to-administration (B-to-A) is a further participant model, which because of its still limited business importance, is not shown explicitly in Figure 6-5. Business-to-administration involves business relationships that can be handled between (retailing) companies and public authorities using an EC system. These include, for example, obtaining import/export approvals, applying for and obtaining stand permits or the goods procurement from authorities over the Internet.

6.3.2 Disintermediation and Reintermediation

Following the presentation of the basic participant model of the EC, we now discuss possible forms of the disintermediation and reintermediation for distribution chains that result from establishing Internet sales and distribution. When the distribution chain between the manufacturer of goods and the end-consumer is considered, it becomes apparent that various market participants can establish an EC interface. The following options form the extreme cases:

- the EC system is installed with unchanged distribution structure at the last middleman, or
- the product manufacturer itself establishes an appropriate system to handle the sales; this can cause the existing distributors to leave the market (disintermediation).

However, because the complete sales handling over the Internet is possible only for digital goods, such as software or media products, examples can be found in practice in which companies continue to use different business models to market their products both over the Internet and over the physical distribution chain. Figure 6-7 shows possible distribution structures.

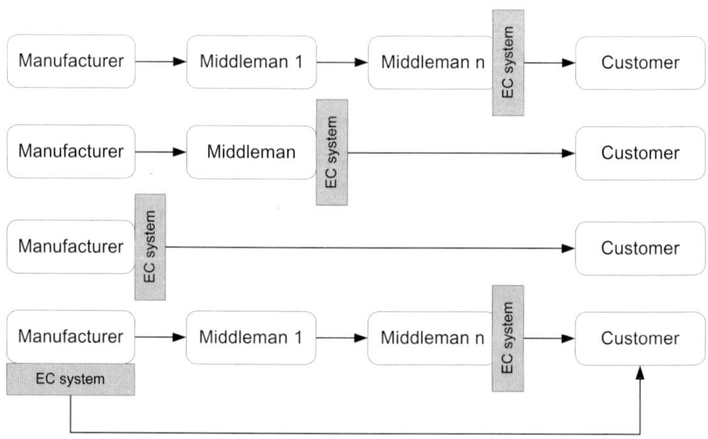

Figure 6-7: EC distribution structures

Multi-stage sales and distribution

The existing sales organization is retained for the multi-stage sales and distribution; the last intermediary implements the EC system here. If, however, the installed DP system is only an isolated measure of the last retailer and does not have any connection to the IT infrastructure of the previous intermediaries, the achievable rationalization effects (costs, time) remain suboptimum. The Internet pages of SURFHAUS (www.surfhaus.de) or Lieferdienst Bünder (www. lieferdienst-buender.de) are examples for the alternative sales organizations.

A model in which a cyber intermediary (reintermediation) replaces the eliminated intermediary represents a variation of this sales structure. The online bookstore Amazon is again a familiar example of such a cyber retailer. Amazon thanks its success in particular to its ability to use the potentials of the information technology and the Internet to provide a significant increase of the up-front customer benefit, during the buying process or on completion. The distribution chain in comparison to a local bookstore remains almost unchanged (see Figure 6-8).

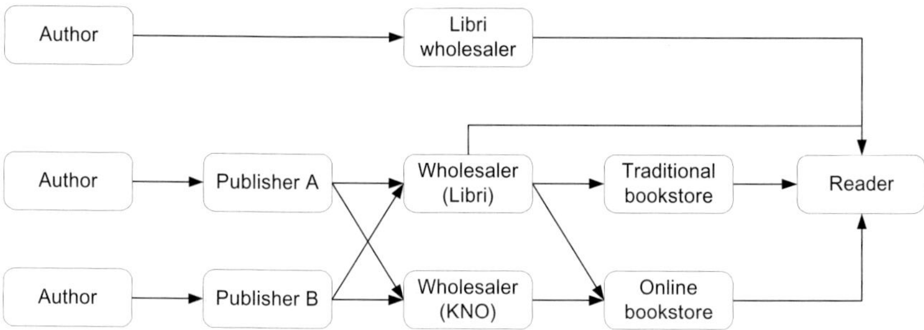

Figure 6-8: Distribution structure in book retailing

Single-stage sales and distribution

When a wholesaler establishes an EC system, the danger arises that downstream retailers in the distribution channel will be eliminated from the market. The lower distribution depth in this case produces a closer integration of the EC system of the intermediary and the DP system of the producer, and thus benefits from a significantly higher rationalization potential. The reduction of the path length of the information flow from the customers to the manufacturer also allows the possibility to more precisely record and document the customers' requirements, and so respond more accurately. The Internet presence of the book wholesaler Georg Lingenbrink (Libri) can be mentioned as example of a distribution structure in which a wholesaler provides the interface to the customer and an EC system to avoid subsequent intermediaries (see Figure 6-8).

The continuing existence of many local bookstores is supported in Germany only by regulatory measures of the state (book price fixing).

Direct sales

In the case of the direct sales, the establishing of an EC system by the producer completely replaces the distribution channel using distributors. This enables the manufacturer to further intensify the communication with the customer and to increase the producer's profit margin. However, the manufacturer then must also undertake the logistics to the customer and is confronted with service, complaints and other questions directly concerned with the immediate interaction with the end-consumer. However, the extent to which the higher margins resulting from the avoidance of distributors are passed on to the customer has not yet been investigated adequately. One of the best-known companies that sells its products exclusively over the Internet is the American company Dell (www.dell.com), which increased its sales from $3.5 billion in 1994 to $18.2 billion in 1998 using this sales and distribution system.

Dual sales and distribution

Historically grown sales structures cause many companies to consider establishing a dual sales and distribution. A corresponding strategy, for example, was favored by Compaq Germany at the beginning of 1999. This envisaged the development of its own EC site for large customers in addition to the EC system for distributors ("Compaq Partner OnlineShops") provided by Deutsche Telekom (T-Mart). However, at the end of 1999, Compaq decided with its "Channels for the Internet Age" sales concept starting in the third quarter of 2000 to sell the majority of its products directly to private and end customers, and so avoid the distribution channel. Here, a logistics partner undertakes the customer-specific configuration and the delivery of the systems.

Companies with products that are very specific and thus have only limited suitability for selling over the Internet continue to be bound to the existing physical distribution channel. To avoid endangering these distribution channels, the possibilities offered by the Internet for the restructuring of selling are only being adopted slowly. This class of companies includes Nike and Puma. Whereas Nike after a long test phase now offers a selection of its products over the Internet (www.nike.com), Puma (www.puma.de) uses the Internet only as an information and communications instrument.

6.3.3 Development Tendencies

The diffusion of existing organizational structures and technologies will further reduce the barriers for market entry in the future and simplify the competitors' access to existing or new markets. The resulting more intensive price competition and the increasing customer demands mean that retailing companies must redefine their role as distributor.

Thus, it no longer suffices to define the service offering of a distributor just with the product palette but with the service package used to position in the market. Possible determinants for the success of a corresponding offering lie in the capability of the retailer to adapt to new technologies and to supplement these with innovative service characteristics. Only the combination of these two factors and their realization in new business models and competition strategies offer an effective means of distinguishing from companies that strive for market success exclusively by adopting technologies (adaptors) or through the development of new services (innovators). Amazon.com is used here as example of the two options.

A business model currently subject to controversial discussion is the "freeware" concept that envisages the "free" provision of products (e.g., computer or mobile telephones). The offering is then refinanced through the provision of services that the customer is obliged to accept. A variation of this model is the "near-freeware" offering from Amazon that offers new customers books for just one US cent. These offerings are refinanced exclusively through the shipping fee or advertising income.

In contrast to the view of a singular market-oriented competitive strategy (i.e., pursuing a strategy of cost leadership *or* differentiation *or* concentration) predominating in the academic world and in practice, Amazon has achieved a pluralistic company strategy:

- With price discounts of between 20 and 50 percent, Amazon.com has assumed the *cost leadership* in US online retailing and has forced its competitors (Books, and Barnes and Noble) to continually enter new price wars.
- Amazon also pursues a *concentration strategy* in which it uses just the Internet as distribution channel.
- With its comprehensive range of supplementary services (gift service, gift vouchers for companies and private persons, personality fan clubs, reader reviews, free music offerings or a personalizable message service), Amazon pursues a *differentiation strategy* in which it tries to satisfy requirements of particular importance for its customers. In this way it aims at distinguishing itself from its competitors and thus achieve a market uniqueness with its offerings.

The time factor must be taken into consideration as decisive quantity to achieve competitive advantages in the emerging electronic markets and to establish information value chains. As Bertelsmann showed, an engagement that is too late can be compensated only at a high financial cost.

6.4 Business Aspects

The automation of the goods, money and information flow along the process chain between provider and customer represents an important prerequisite for economic success. The many media fragmentations associated with the existing information transmission (fax, telephone, e-mail and post) mean that the customer's central success determinants (where, when or how the service is to be provided) are not fully satisfied (see Figure 6-9).

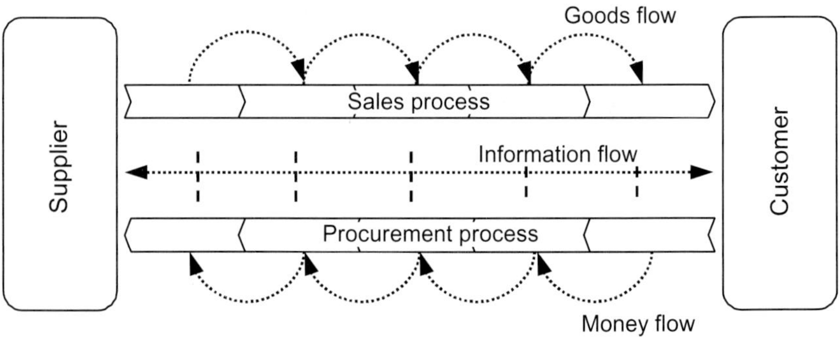

Figure 6-9: Interfaces in the process handling

The use of EC applications that also integrate the participants' retailing information systems or enterprise resource planning systems (ERP systems) allow a reduction of the information flow path both for the transmission and during the provision of the service, the continuous monitoring of the service handling (delivery, payment), and, if necessary, initiate corrective measures in real time.

The following section uses an example of the procurement and sales process to illustrate the new capabilities of the service handling provided by EC. To emphasize the differences to the traditional process handling, the participant model, which offers the highest development potential, is used here. Because the two processes represent mirror images, after considering the electronically-supported procurement process (electronic procurement), the B-to-B model is used as example to discuss the sales process as part of the B-to-B-to-C model.

6.4.1 Electronic Procurement

Whereas the procurement process for goods in retailing companies could be optimized early through the use of retail information systems in combination with EDI, the procurement process for goods used to maintain the operational function (maintenance, repair and operations) still exhibits many deficits. These include:

- an inefficient selection of products and providers
- time-consuming, manual processing of requirements and purchase orders
- expensive and unnecessary warehouse storage
- long approval procedures
- unsatisfactory control capabilities, and
- process costs that, to some extent, significantly exceed the product costs.

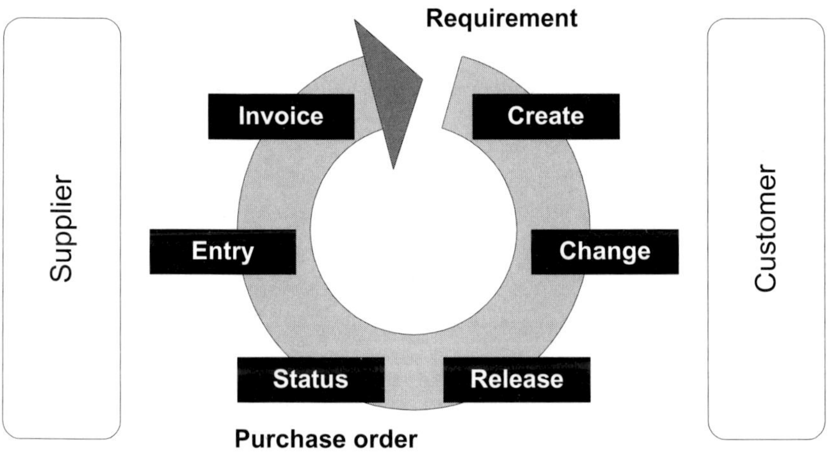

Figure 6-10: Low-interface process handling using electronic procurement

The advantages for the company that result from the use of an electronic procurement application (see Figure 6-10) can be classified with the dimensions time saving, cost saving, increase of the procurement quality and strategic success potentials (Nenninger et al. 1999, p. 287).

Time saving:
- faster handling of purchase orders
- reduction of the times between the order receiving and the dispatching, and the delivery times
- reduction of the processing times in the correspondence with authorities (customs papers, import and export approvals)

Cost saving:
- reduction of the personnel costs
- reduction of the shipping and storage costs

Increase of the procurement quality:
- improved meeting of deadlines
- reduction of data entry errors
- new capabilities of procurement marketing

Strategic success potentials:
- better binding of vendors
- transfer the operative handling to the requirements requester
- improvement of the planning accuracy

Section 6.6.2 uses the example of the SAP business-to-business procurement solution to introduce the functional form of a corresponding application.

6.4.2 Electronic Selling

EC applications allow providers to handle a larger number of customers online while addressing them personally. However, the increased market transparency and the increasing customer demands place increasing requirements on the functionality of the systems.

The example of a four-phase transaction model (Figure 6-11) is used to explain the handling of a sales process:

The product presentation by the provider and the market observation by the customer are performed in the *information phase*. The customer can analyze the general, industry and technology overall business information to obtain an overview of the market. The customer can also obtain information about potential business partners (e.g., with regard to solvency or delivery capability), and their products and services (e.g., with regard to specifications or conditions). The electronic product catalogs that display the service offering provide not only the presentation and ordering function, but increasingly also functionality from retail information systems, such as calculation or availability control, and interfaces to the corresponding systems.

The *agreement phase*, in which the conditions for the completion of a contract are negotiated, follows the information phase. On completion of the agreement phase, the market participants agree to exchange services. However, because of the fixed prices applying in the private customer area, negotiations seldom occur in this sub-transaction. Agent-supported systems that allow customers to compare offers and possibly request the system to obtain a better offer are an exception here. Auction systems represent a further possibility of dynamic pricing that, for example, can support the retailer in selling special items for seasonal goods.

Digital signatures can be used to guarantee the authenticity of the documents exchanged in this phase (quotations or purchase orders). In Germany, the "Law for regulating the general conditions for information and communications services" from June 13, 1997 provides the legal basis here. The EU regulations for electronic signatures have taken effect with their publication at the beginning of 2000 (EU 2000). They require that member states give a non-counterfeitable signature the same legal status as a handwritten signature.

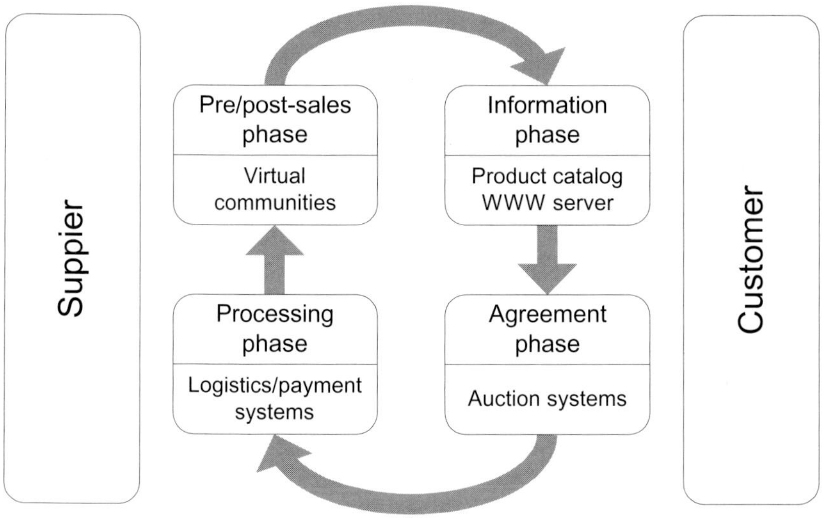

Figure 6-11: Four-phase transaction model

The service agreements, such as packing, transport or picking of the goods, are satisfied in the following *processing phase*. Logistics support systems and payment systems are two services used in this part of the transaction processing. In the meantime, all large logistics companies offer a corresponding interface to the Internet and so provide their customers with the possibility to track online the processing progress within the logistics chain and also the option to subsequently check the progress of the provided service.

Currently, none of the Internet solutions developed for payment processing (CyberCash, eCash, MilliCent, Geldkarte, Verifone, etc.) has been able to establish itself as market leader. However, in addition to the dominating traditional procedures for payment processing with invoice, cash on delivery or cash payment on receipt of the goods, an increasing acceptance of the use of credit cards is becoming apparent. The SSL protocol or the SET (Secure Electronic Transaction) standard is used here.

Whereas SSL supports only the secure transmission of the credit card information to the retailer with optional client authentication, SET guarantees a secure processing of the complete payment process. To ensure the identity of the transaction participant, SET assumes the use of digital certificates both at the provider and the customer side. SET-supported payment transactions need a SET certificate and a SET wallet at the customer location, and a SET-conform server and a SET certificate at the retailer. Currently, 15 banks in Germany support the SET Standard (www.visa.de).

The *pre/postsales phase* covers all measures before and after the purchase that serve to bind the customer stronger to the company and so induce the customer to buy additional products. The applications used here can be described using terms such as "electronic customer care" or "customer relationship" tools. They are used to display information and the communication not only between system operators and customers, but also between customers. The solutions that support the associated measures for customer loyalty include push/pull news services, chat forums, product configurators and applications to establish virtual communities.

6.5 Technological Concepts of the SAP Systems

6.5.1 SAP IDocs

In addition to the option of the synchronous data exchange, the possibility of asynchronous data transmission is critical for EC applications. The *Intermediate Document* (IDoc) interface defined by SAP, which allows users, in addition to the data transmission using ALE (see Section 4.6), also to transfer data between applications, uses EDI standards (ANSI X.12, EDIFACT) to exchange business messages. Thus, this concept provides the capability,

- to structure and automate the document exchange,
- to reduce the structure complexity of different application systems through the use of standardized document types (e.g., for purchase orders), and
- to define a container for data that from the application viewpoint are inconsistent and so must be processed separately.

The data exchange between the involved application systems is performed in three steps:

The *initial processing* covers the generation of the data to be transferred by the SAP application, their transfer to the IDoc interface (direct or using the message control) and the forwarding of the IDocs to the external system (e.g., an EDI subsystem). If the second system is connected to the SAP system through the file interface, it is also possible to test whether the system can be started automatically.

The *input processing* covers the acceptance of the IDocs, their syntactical verification and the starting of the appropriate SAP business process.

The *status processing* indicates the measures that the external system initiates from the sending system as result of the transmission of a control report. The type of the exception handling is determined by the value of the status records in the transmitted status report.

6.5.2 SAP Business Objects

A *business object* (BO) is the information-related representation of a business entity (such as vendor, customer or sales order). Thus, it permits the user to invoke an encapsulated software object to access a specific business data basis. A BO is built using a four-layer shell model (see Figure 6-12):

- the object data form the *kernel* of an SAP business object
- the second shell, the *integrity layer*, covers business rules and general conditions, and represents the business logic of the object
- the third shell, the *interface layer*, defines the interfaces of the object to the outer world
- the fourth shell, the *access layer*, defines the technologies used to access the objects.

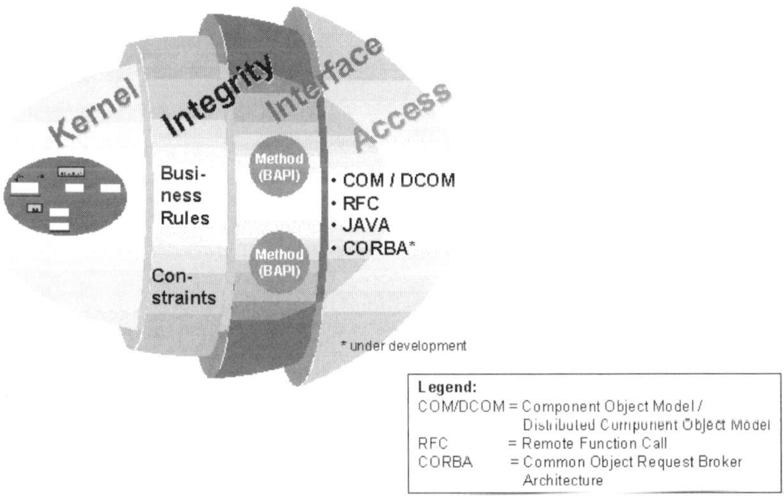

Figure 6-12: Shell model of the SAP business objects

The properties that characterize an object include the events (that can cause a state change of the object), the attributes (which describe the data stored in an object) and the methods (which characterize the operations used to process a BO). The assignment of the methods – of the BAPIs (*Business Application Programming Interfaces*) – to a BO is performed in the *Business Object Repository*. The BAPI browser displays all business objects implemented for the BAPIs.

The advantages for the user that this technology provides include:

- the decoupling of the business view from technological developments
- a high interoperability through the possible use of different interface technologies
- a reduction of the development costs and a faster implementation through the reusability of objects.

BAPIs can be invoked by *Internet Application Components* (IACs) using remote function calls (RFCs) or the DIAG (Dynamic Information and Action Gateway) protocol.

6.5.3 Internet Transaction Server and Internet Application Components

Confronted with the increasing commercial use of the Internet, SAP also saw the need in 1995 to develop concepts for the use of SAP systems in the Internet. The *Internet Transaction Server* (ITS) introduced a means of using SAP applications (transactions) over the Internet. This section presents the most important elements of the ITS (see Figure 6-13).

Web-Gateway (W-Gate)

The W-Gate assumes an intermediary role between the Web server and the application gateway (A-Gate). It uses an interface to forward the requested data to the A-Gate, from which it then accepts the prepared HTML pages. It supports the CGI (Common Gateway Interface), NSAPI (Netscape Server Application Programming Interface) and ISAPI (Internet Server Application Programming Interface). The interface is separated into two components (W-gate and A-gate) primarily to minimize the security risk.

Application-Gateway (A-Gate)

The A-Gate forms the link between the W-Gate and the SAP system. It manages the user sessions and the system connections. It also performs the representation of the SAP data on HTML pages and its multithread capability allows the simultaneous processing of multiple queries.

Service file

The service description contains details that allow the A-Gate to establish a connection to a dedicated SAP system. These data include the transaction name and login information, such as user name, password and login language.

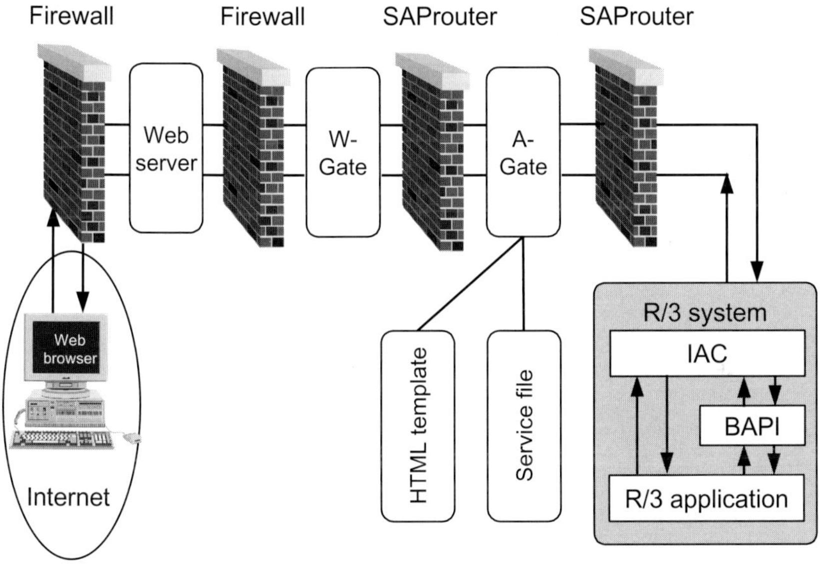

Figure 6-13: Connection of an SAP system to the Internet using the ITS

HTML template

Templates are predefined patterns that represent the SAP data in the Internet. In addition to the standard HTML code, these contain HTML^{Business} instructions that allow a browser to represent the objects displayed in the WINGUI (previously SAPGUI; see Section 4-3).

Security components

As shown in Figure 6-13, a multi-level security concept can be used to ensure the security of the data saved in the SAP system. In addition to the option of using firewalls and SAProuter, it is also possible to encrypt the communication between the browsers and the Web server using SSL. From Release 4.5B, ITS also supports X.509 certificates for user authentication.

SAProuter

SAProuter is a program supplied as standard with the SAP system that should be installed as supplement to a firewall in order to restrict application-level access to the SAP system. The use of SAProuter allows additional protective mechanisms to be realized prior to the access, such as for authentication or access restriction.

Internet Application Components (IAC)

Internet Application Components designate prewritten applications that consist of HTML templates, service descriptions and R/3 or Retail transactions. In addition to the possible use of the Internet as access channel, these applications differ from the standard transactions of the SAP system mainly in the reduction of the application complexity and a simple operation through the intuitive Web user interface. Table 6-3 lists a selection of IACs that are useful for the development of Internet applications in the retail sector. The table shows the designation, the SAP module in which it is used, and the release version when it was implemented.

Table 6-3: Available IACs

IAC	Module/Application	Release
Product Catalog	Logistic (SD/MM)	(3.1G)
Available-to-Promise	Logistic (SD/MM)	(3.1G)
Sales Order Creation	Logistic (SD/MM)	(3.1G)
Sales Order Status	Logistic (SD/MM)	(3.1G)
Quality Certificates	Logistic (SD/MM)	(3.1G)
Consignment Stocks Status	Logistic (SD/MM)	(3.1G)
Service Notification	Logistic (SD/MM)	(3.1G)
Requirement Request Status	Logistic (SD/MM)	(3.1G)
Collective Release of Purchase Request	Logistic (SD/MM)	(3.1G)
Collective Release of Purchase Orders	Logistic (SD/MM)	(3.1G)
Project Data Confirmation	Logistic (SD/MM)	(3.1G)
Kanban	Logistic (SD/MM)	(3.1G)
Project documents	Logistic (SD/MM)	(4.0A)
Procurement via catalogs	Logistic (SD/MM)	(4.0A)
Status information	Logistic (SD/MM)	(4.0A)
Who is who	Employee Self Services (HR)	(3.1G)
Employment opportunities	Employee Self Services (HR)	(3.1G)
Application status	Employee Self Services (HR)	(3.1G)
Calendar of events	Employee Self Services (HR)	(3.1G)
Booked events – R/3 users	Employee Self Services (HR)	(3.1G)
Booking attendance (R/3 users)	Employee Self Services (HR)	(3.1H)
Booking attendance (Web users)	Employee Self Services (HR)	(4.0A)
Cancel attendance (R/3 users)	Employee Self Services (HR)	(4.0A)
Cancel attendance (Web users)	Employee Self Services (HR)	(4.0A)
Address information	Employee Self Services (HR)	(4.5A)
Family member information	Employee Self Services (HR)	(4.5A)

Emergency address information	Employee Self Services (HR)	(4.5A)
Emergency contact information	Employee Self Services (HR)	(4.5A)
Bank information	Employee Self Services (HR)	(4.5A)
Hire event	Employee Self Services (HR)	(4.5A)
Customer account information	Financial (FI)	(3.1G)
Asset information	Financial (FI)	(3.1G)
Internal activity allocation	Financial (FI)	(3.1G)
Internal price list	Financial (FI)	(3.1G)
Invoice verification	Financial (FI)	(4.5A)
Display record trips (FI)	Financial (FI)	(4.5A)
Load trip costs recording (FI)	Financial (FI)	(4.5A)
Stock allocation	SAP Retail Store	(4.5A)
Promotion	SAP Retail Store	(4.5A)
Consumer order	SAP Retail Store	(4.5A)
Consumer order status	SAP Retail Store	(4.5A)
Physical inventory	SAP Retail Store	(4.5A)
Goods movement	SAP Retail Store	(4.5A)
Goods receipt	SAP Retail Store	(4.5A)
Store order	SAP Retail Store	(4.5A)
Assortment list	SAP Retail Store	(4.5A)

The more than 85 IACs currently supplied by SAP can be assigned to various modules. However, the development of dedicated EC applications (see Section 6.6) means that these will replace some of the IACs made available with Release 3.1 G (for example, "Collective Release of Purchase Requisitions" or "Collective Release of Purchase Orders"). Such IACs will not be further developed.

6.6 SAP Internet Applications

SAP has combined some of the previously developed IACs to produce the EC applications: SAP Online Store, SAP Business to Business Procurement and SAP Employee Self-Service.

6.6.1 Online Stores

The sales of products over the Internet can be made either using the SAP Online Store (see Section 4.5) or with special EC software from third-parties. However, to satisfy a company-wide information and data management, the third-party solutions must fulfil a number of requirements. These include:

- a consistent representation of the company information flows using transaction processing
- real-time provision of information
- full support of business processes
- the guarantee of transactional integrity
- fault tolerance for component failure through asynchronism.

When the available interfaces (for example, RFC, EDI or ALE) are evaluated taking account of the requirements, it is apparent that a coupling using ALE is the best variant.

The Connectivity Software from SQL GmbH (www.sql-gmbh.de) is a software solution that follows this concept. Its characteristics are listed below.

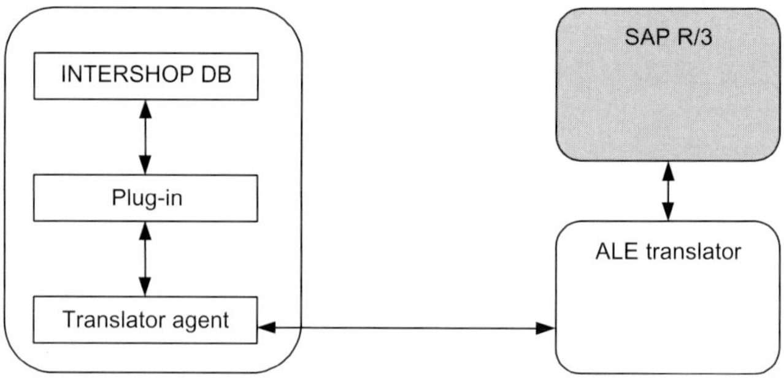

Figure 6-14: Connection of an eShop to SAP R/3

The Connectivity Software consists of three components (see Figure 6-14):

- a *plug-in* integrated in the database of the eShop software is responsible for the bidirectional data transfer
- the *ALE translator* that processes the application data combined as IDocs
- a *translator agent* that provides the data transmission between the translator plug-in and the ALE translator.

The plug-in and the translator agent are installed on the computer on which the eShop application is implemented. The ALE translator is installed on a computer in the LAN of the R/3 system. The R/3 ALE settings must be customized to be able to successfully exchange data with the SAP system using the ALE interface. To avoid delays in the data matching between the SAP and the eShop system, the network section between the SAP application and the ALE translator should be as short as possible.

The plug-in registers every new purchase order entered in the eShop. It then collects all the data needed to process the purchase order and makes these data available to the translator agent. This returns the data via network to the ALE translator. The booking is made automatically once the ALE translator has passed the purchase order to the SAP system. Once the purchase order has been processed and stored by the SAP system, the order number is automatically reported back to the ALE translator. The eShop database plug-in stores the order number and is available for the subsequent processing.

The INTERSHOP 3/4 and INTERNOLIX Business Line software products are two eShop systems that can be connected to SAP R/3 using this concept.

6.6.2 Business to Business Procurement

SAP Business-to-Business Procurement was introduced with Version 1.0A in 1998 and has release cycles independent of the SAP kernel system. It has its own catalog that manages a preselection of products that are of importance for company employees as part of the procurement of B- and C-goods. There is also the option to access external catalogs (from manufacturers or brokers) or to integrate these catalogs in the internal intranet. The embedding of the SAP Business Workflow also provides the capability to automate procurement processes. For example, a purchase order request that exceeds a specified value can be forwarded automatically to the supervisor. If he makes a positive decision, the person placing the order is informed by e-mail or work item, the purchase order request is posted in the company system and forwarded to the vendor. The following section discusses some of the available functions.

Create user master records

The use of templates means employees that wish to create their own data master record only need to enter their personal details (name, communications data, etc.); the system derives the additional business data. The employees can themselves update their master data.

Create requirement request

The shopping basket provided to perform the requirement request can have a default assignment corresponding to the individual buying profiles. Basic data, accounting, notes, availability and sources of supply are displayed as details.

Process requirement request

The requirement request can be processed as far as generating the requirement document in the ERP system.

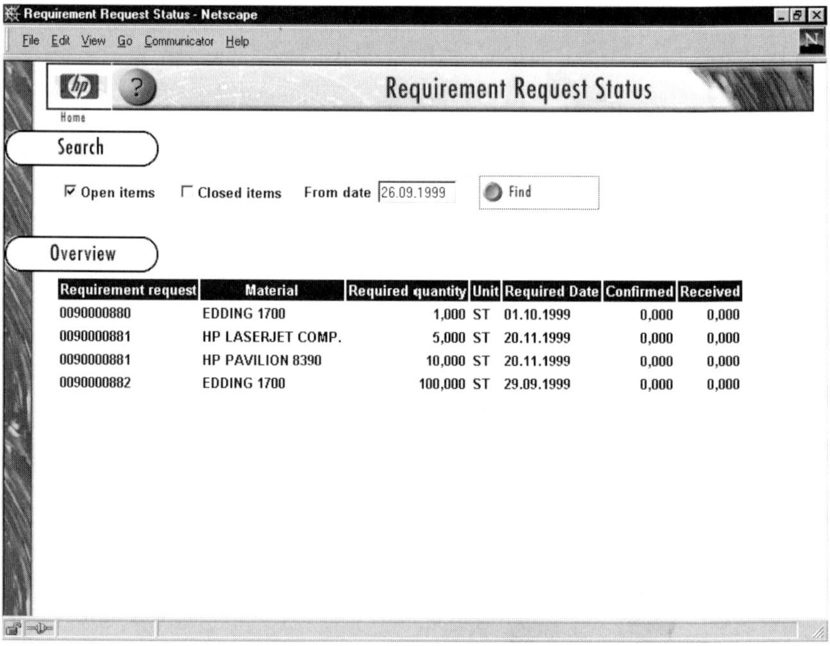

Figure 6-15: Display of a status check

Test the status of the requirement request

The requester can continually check the progress in the processing of the requirement request. This allows him to determine whether requirements have been approved or rejected or are incorrect, and, provided a connection exists to the DP system of the carrier, track the processing progress along the logistics chain (see Figure 6-15).

Confirm goods receipt

An employee confirms the acceptance of the ordered goods. If the delivery is incomplete or damaged, the clerk can make appropriate comments or return the delivery.

Approve invoice

Once the sales order has been processed, the vendors can initiate the corresponding invoices. After the invoice has been posted, it is presented to the requestor and released for payment after approval. A Web-based workflow inbox or the e-mail clients are used to support the automation of the approval procedure.

Analysis of the purchase orders

Provided they have the appropriate authority, decision makers can analyze all purchase orders present in the SAP system. The cost center, vendor and the requestor can be used as selection criteria.

Release 2.0A of the Business-to-Business Procurement application released by SAP in spring 2000 also provides the capability to integrate third-party ERP systems.

6.6.3 Employee Self Service

The EC application allows employees with the appropriate authority to manage a selection of their own personal data. In addition to the increase of the data quality by shortening the time window needed for the update, this application also helps to reduce the administrative costs in the personnel area. The use of such a system is particularly appropriate for retailing companies where relatively many employees have access to a computer system, such as at the company headquarters. The following section discusses some of the functions provided for the various tasks.

Process person-related data

Authorized employees are provided with the capability to update their contact information (private address, telephone number, etc.) and leave details for emergencies. Information about tax class, tax allowances or bank account can be changed.

Time recording

Employees can use the application to enter work times, assign these to the different projects and obtain a time accounting. An option is also provided to apply for vacation, and to obtain an overview of the previously approved vacations and vacation still free.

Reserving of advanced training seminars

The application provides the user with attributes such as topic, language or training location to select those offerings from an overview that are of interest (see Figure 6-16). Not only registration and cancellation are offered, but also the option for making an advance reservation for seminars for which a date has not yet been set.

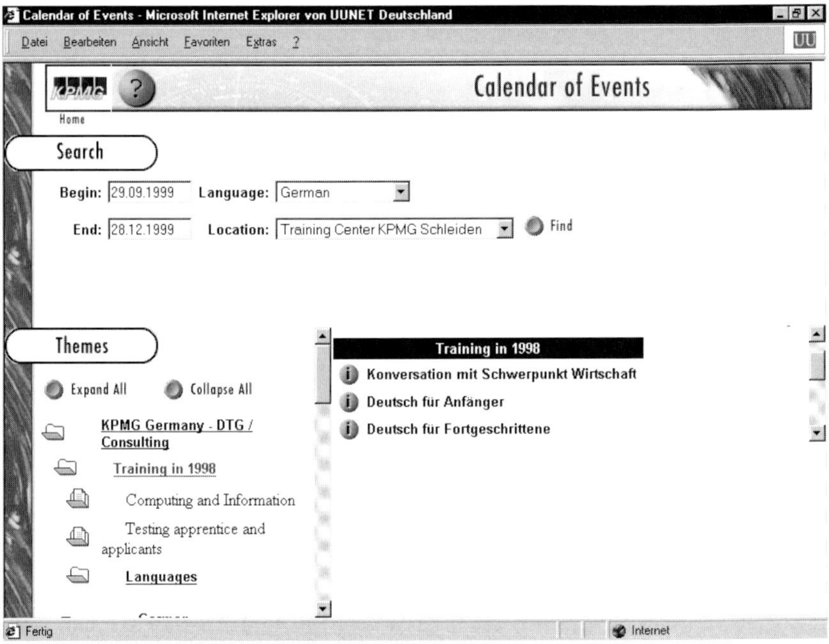

Figure 6-16: Display of the possible seminars

Person directory

The Internet/intranet provides the possibility to obtain employee-specific information such as name, workplace, communications data or the hierarchical position.

Further functions are provided for the administration of the travel costs and expenses. Queries can also be performed for employment and salary statements or payroll accounting.

7 Planning and Control of Goods Flows Between Industry and the Retail Sector

7.1 Efficient Consumer Response as Business Concept

7.1.1 Terms and Goal

The relationship between industry and the retail sector has traditionally provided a high conflict potential. This results primarily from the different goals pursued for the policies for product and price on one hand, and communications and distribution on the other hand. As part of the product and price policy, manufacturers are mainly interested in building a positive image of their own brands and products. In contrast, the retail sector favors providing a positive image of its assortment and shopping outlet. This includes the different emphasis on manufacturing and retail brands. As part of the distribution policy, industrial companies tend to favor the delivery of large order quantities, whereas the retail sector wishes the fastest possible delivery, even for small volumes. Advertising by the manufacturer in the conflict area of end-consumer and permanent promotions by the retail sector serve to devalue brands. Increasingly, industry and the retail sector are faced with the need to meet joint profit goals, recognize this unsatisfactory situation as chance to end the conflicting relationship and to use cooperatively all success potentials available within the value chain to the end customer (see Figure 7-1). Not aggressive low prices, but a better price-benefit ratio, should produce more customer satisfaction and thus larger market shares and increased sales.

Efficient Consumer Response (ECR) designates a still relatively new business concept to improve the relationships between industry and the retail sector. Consequently, the scientific literature does not yet provide standard terminology for this subject. The ECR term is generally used to describe the bundling of various strategies for the optimization of the value chain, to maximize the customer satisfaction and to achieve synergy effects as cooperation between industry and retailing based on mutual trust. ECR Europe defines Efficient Consumer Response as "... *a strategic initiative working to overcome traditional barriers between trading partners, thus eliminating internal barriers that result in costs and time that add little or no value to consumers. Efficient Consumer Response is focused on the application of leading edge management methods and available technologies to reduce costs and response times, while increasing the quality of products and services that are provided to consumers*" (ECR 1999).

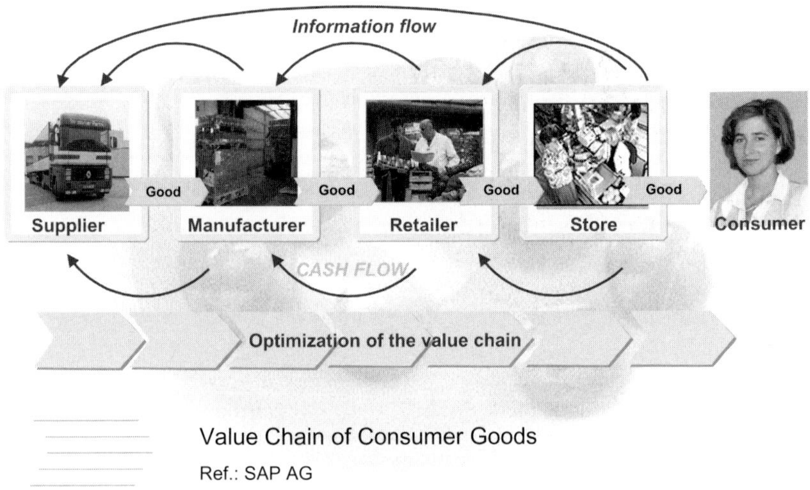

Figure 7-1: Value chain through to the consumer

A fast and precise response to customer demand is generally conceivable even without changes being made to the value chain: An increase of the stockholding can give manufacturers and retail companies increased flexibility in responding to changed customer requirements. However, the resulting increased storage costs affect the prices. High warehouse stock levels are a characteristic property of the *push principle* that dominates in conventional value chains. Here, the manufacturers attempt to sell their products in large numbers and as independently as possible from the actual demand of the consumers. The principle tool here are the conditions granted by the manufacturers to induce retailing companies to accept larger quantities for warehousing. Cooperation between the process levels of the value chain takes place in isolation, if at all. Thus, only suboptimum solutions are possible. ECR represents a general trend to reverse this principle to achieve a *pull principle*. Here, the needs and wishes of the customers form the starting point of the goods flow along the value chain. Requirements data based on the recorded sales provide the manufacturers' planning base. Thus, the replenishment supply of products to retailers is continuous and oriented on the actual requirements of the consumers. The cooperation and information flows that exist between the individual levels of the value chain or supply chain permit a process optimization over distribution levels. Normally, push and pull designate purchasing strategies within a company. These terms are used here to explain the trend that the goods supply to the market is not primarily geared to the production capacities of the manufacturer, but is principally oriented on the demand situation. The transition from the push principle to the pull principle should be interpreted as being a trend statement. Depending on the situation (also refer to Section 7.2.1), the two principles complement each other.

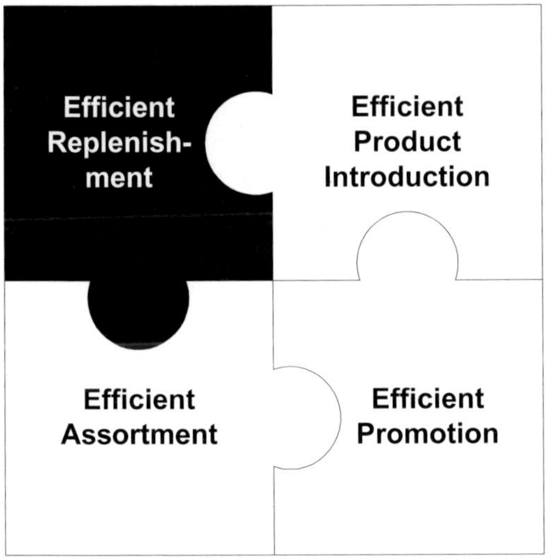

Figure 7-2: The strategy mix of the Efficient Consumer Response

ECR contains four basic strategies of a partnership-oriented cooperation between manufacturers and retailing companies (Figure 7-2) that should provide additional value for the consumer coupled with lower costs for manufacturers and retailers as result of the inter-company control of the goods and information flow along the supply chain and the cooperative marketing:

- **Efficient Replenishment (ER)** or Continuous Replenishment: connects modern logistical concepts along the supply chain with automated ordering system and electronic data exchange.
- **Efficient Assortment (EA):** aims at improving the assortment productivity using shelf and shop area optimization (space management).
- **Efficient Promotion (EP):** includes the planning and control of appropriate sales promotion activities; it aims at improving the efficiency of the communications policy along the value chain.

- **Efficient Product Introduction (EPI):** designates the close cooperation between manufacturer and retailing company for the development and introduction of new products; it aims at cost savings, gaining time and reduction of the flop rate.

Whereas ER is the logistics-oriented component for ECR, the other base strategies are marketing-oriented. Although the strategies can also be pursued in isolation, they attain their full effect only in combination.

7.1.2 Origins and Development

The cooperative concept of ECR is not basically new. Isolated efforts by manufacturers in order to improve the binding of the retail sector to the own company can be considered as conceptional forerunner of this concept. An early form of this cooperation is the *key account management*. Here, specific retail customers particularly important for the market success of the company are handled separately as key accounts. From the industry viewpoint, the customer manager adopts a central role in this field. The customer manager's tasks include the collection and preparation of customer information, the customer-related business planning, the preparation and negotiation of agreements with the customer, and the control of the degree to which the customer's goal is achieved. The customer manager remains as contact partner for queries, complaints or other contact wishes by the customers. A further development of the key account management is the *trade marketing* that has the goal to establish certain preferences of the retail sector with the manufacturing company operating the trade marketing. In addition to the conventional consumer marketing based on the strength of the brand, the marketing orientation is on retailing companies.

The *quick response concept*, developed in 1984 in the USA for the analysis and rationalization of the value chain in the textile and clothing industry, starting with the fiber production and ending with the sales of ready-made garments, can serve as direct predecessor of the ECR concept. It represents an application of the previously mentioned pull principle in which the customer requirements form the starting point for all earlier activities. The prerequisite for a functioning quick response is the early recognition and the fast meeting of customer demands. The following scenario can be used to illustrate quick response during the course of a season: clothing manufacturers retrieve daily sales data for their goods and those of the aggregated competitors from the retail companies or their head offices. With these data, the manufacturer can judge the business trend, make early purchases and possibly initiate the preproduction and direct revision of the assortment for the retailing company (refer to the discussions for the Vendor Managed Inventory in Section 7.1.3). In return, retailers can view quantity and type of the available goods in the warehouse, production and purchasing at the clothing manufacturer, and, when necessary, reorder directly. Consequently, characteristic for quick response is a relatively high number of demand-oriented – and thus reduced-risk – reorders coupled with reduced preorders at the start of the season. The communication between the clothing manufacturer and the material vendors is similar, however, with data aggregated at a higher level. The data for the order and purchasing trend at the clothing manufacturer allow the material vendors an early and continuous purchasing and possibly production for stock. In turn, the clothing manufacturer has the possibility to retrieve important information about production for stock, purchasing and stocks with exact specifications, such as material widths, color shades and quality characteristics.

The management consultants Kurt Salmon Associates (KSA) and the Food Marketing Institute initiated the first ECR projects in 1992 in the USA. The base technologies needed here, in particular scanning and electronic data exchange, existed already. ECR initially spread relatively quickly in the USA. ECR Europe, founded in 1994 as joint organization from industry and retailing, had the task, in close cooperation with national ECR initiatives, to promote projects at the European and national level. Information about ECR Europe can be found in the Internet under http://www.ecrnet.org/ECR/ECR.home.

7.1.3 Efficient Replenishment

Efficient Replenishment (ER) designates the logistics-oriented strategies for ECR. Logistics is understood to be the planning, control, realization and monitoring of the goods flows both within and between companies, and the accompanying information flows. In the retail sector, logistics applies in particular to the areas of transportation and stockholding. The disposal logistics relating to the residual material flows is also gaining importance for retailing companies. However, this is not to be considered in further detail here. As trend, the retail sector is pursuing a reduction of its stock levels. This implies more frequent deliveries in smaller quantities. Nevertheless, a high goods presence should be maintained in the sales area. Consequently, high demands are placed on the logistics along the supply chain. Inefficiencies should be avoided by matching the control of the flows for goods and information (*supply chain management*). The process procedures that replace traditional ordering and delivery are supported by inter-company integrated information systems that control the supply of the retail sector based on the actual sales or sales forecast using an appropriate procedure. The goals are the lowering of stock levels and the reduction of the duration of goods in the chain, i.e., to accelerate the goods flows. In addition to industry and retailing, in particular, logistics service providers are included as partners. The *ER base technologies* are internationally standardized identification systems, such as the International Location Numbering (ILN), the article number (EAN) for the identification of logistical shipping units, scanning and EDI, nowadays normally based on the EANCOM standard, a industry-specific subset of EDIFACT for the consumer goods industry and retailing (see Section 6.2.1).

Computer-supported purchasing systems that analyze scanner data use forecasting procedures and automatically generate order proposals or purchase orders or release deliveries are a prerequisite for ER (see Section 3.3.2). The responsibility for the ordering system can be controlled in different ways. It can lie exclusively with the retail company, the manufacturer can be included in the process, for example, by transferring order proposals, or it can be passed completely to the manufacturer. The latter concept, designated as *Vendor Managed Inventory (VMI)*, best meets the basic philosophy of ER. In VMI, the decisions for the delivery cycle and the delivery quantities are not made in the retail company but by the vendor. Although the retail sector forgoes determining the order quantities and times, it regularly supplies the manufacturing company with sales figures and inventory data. The process runs in the several steps outlined below:

- Retail companies send daily and for each article the stock level, the order level, the sales and the shortfall quantities directly to the manufacturer or to a mailbox.
- The MRP controller at the manufacturer checks these data for plausibility and completeness.
- In regular intervals, e.g. weekly, the MRP controller performs a sales forecast using the transferred data. The data determined in this way enter the manufacturer's planning.
- The specified reordering point and the available stock level are used to determine the delivery quantities after a delivery allocation has been agreed previously between the retailer and the manufacturer.
- Because manufacturers normally deliver to several retail customers, a load planning of the transportation equipment is then performed.
- An order confirmation is sent to the retail company where the goods receipt planning then can be performed.
- The retailer allocates the order number and after receiving the order confirmation sends its own confirmation with the allocated order number back to the vendor.
- The "purchase order" then enters the vendor's order processing.

Both vendors and retail companies share the advantages of this procedure. The manufacturers now have valuable data that allow them to make realistic production planning and to reduce stocks of completed goods. A higher goods presence and improved utilization of the transportation capacity can also be achieved. The major advantages for the retail companies lie in the lower stock levels, reducing the tied-up capital, and in an increased goods presence at the POS. However, the procedure requires a certain degree of trust between manufacturers and retailers.

To avoid compensating for the rationalization advantages of the reduced stock levels with increased costs in the goods receiving, the retail sector is dependent on the bundling of the goods delivery. Consequently, modern warehousing concepts are gaining in importance for ER. Goods flows are bundled in central warehouses and separated at the following stages. Transit terminals and goods distribution centers, whose central concept is designated as *cross-docking*, represent a further development of the central warehouses. Cross-docking is a process within the supply chain in which the supply of goods to the goods distribution center and the delivery to the recipients is coordinated in time and quantity to avoid the stocking and destocking processes (see Section 3.3). Immediately after receipt, the goods are prepared for shipping and transported to the appropriate goods issue points without needing to be repacked. Thus, goods distribution centers realize a pure distribution function of vendor-related inbound deliveries to store-related outbound deliveries. Here, the goods carriers packed with a single variety at the manufacturer are checked for completeness and delivery state during the goods receipt, and assigned to the stores. Cross-docking requires the cross-company integration of the DP systems using the previously mentioned base technologies. This permits the reduction of stock levels and the associated costs, and allows products to be delivered earlier and fresher. It also reduces the area required. The orientation of the order of loading in the goods distribution center to the store layout saves unloading time at the stores. A combination of cross-docking with VMI is advantageous for the vendors. This is equivalent to the manufacturer assuming the control of the distribution centers.

ER is often designated as the central base strategy of ECR (Heydt 1998), namely as the initial strategy on which the marketing-oriented base strategies described in the following sections can build.

7.1.4 Efficient Assortment

The building and the continuing management of an attractive assortment are significant factors for establishing the profile of a retail company. The assortment management means not only responding to developments in the buyer behavior in the past, but, in particular, acting on new trends in product development and in customer demand. The retail sector is primarily interested in optimizing sales and revenue of the complete assortment over a longer time period. In contrast, manufacturers endeavor to increase awareness of their products with retailers compared with those of competitors. The strategy of the Efficient Assortment (EA) acts at this interface between industry and the retail sector. EA is understood to be the efficient *structuring of the assortment* by manufacturers and retail companies with the aim of making optimum use of the space available in the sales areas and of maximizing customer benefits.

Traditionally, the retail sector alone was responsible for the task of the assortment planning, control and monitoring. EA should simplify the realization of these tasks by providing a trust-based cooperation between industry and the retail sector. This provides the sensible combination of the retail-side knowledge for area layout and assortments with the manufacturer-side knowledge about individual products, marketing and results of the market research. The different sales outlets must be taken into consideration here, e.g., clusters of similar stores must be formed. Consideration must be made of the different target groups and also interdependencies of the articles in the assortment, which, for example, can result as effects of combination and cannibalization.

Manufacturers, retail companies and buyers should all profit when the available space in the sales outlets is used in the best possible way. Consequently, the most important methods for EA are the shelf and area optimization (*space management*), and the closely associated methods of assortment controlling, such as direct product profit (DPP) analysis, conjoint analysis or process cost calculation (see Chapter 5). Further components of the EA are used to determine the locations of the merchandise categories and articles, the contact sections and prices. Space Management (SM) is the computer-supported shelf space planning and control, in particular for the assortments of large sales outlets such as supermarkets. SM optimizes the use of the sales area and the shelf space with regard to sales and revenue. SM should always achieve the following goals (Heydt 1998):

- placement of articles appropriate for the sales volume to avoid stock shortages and excess stocks, and to ensure uniform shelf sales;
- achievement of an effective selling shelf appearance by building horizontal and vertical blocks, and ensure a long-term order in the shelf;
- determination of an optimum assortment range and depth through the discontinuation of unprofitable products and augmenting the assortment to provide consumers with an offering that meets their demands while achieving the best possible profit;
- creation and control of the customer demand.

A typical shelf optimization process runs in several phases. After a basic *determination of the general conditions*, in which representative test and control markets, and the desired stock range must be defined, the *acquisition of relevant data for the actual situation* is performed initially. In particular, this involves the dimensions of the shelves, size of the products and their previous shelf placement, sales quantities, and purchase and selling prices. In the following phase, the analysis of the shelf, weaknesses of the previous article placement are determined and visualized using shelf graphics. One of the tasks of the following *optimization of the shelf* is to suggest a new assignment of the articles for a shelf. Here, initially lower and upper limits are defined, such as the number of articles to be presented, the stock duration or the article frontal area. These are then used to determine a new shelf assignment. Optimization criteria can be output-oriented quantities such as sales quantity or the sales value, or cost-oriented criteria. In addition to the gross yield, namely the comparison of business volume and delivered costs of the goods, in particular, the direct product profit (DPP) of an article has attained significant importance as optimization criterion. The placement suggestions determined exclusively using quantitative characteristics can now be adapted using qualitative criteria, for example with company-specific placement strategies. The *realization of the optimization suggestions* completes the shelf optimization processes. After some time, the success of the shelf reorganization is checked and, if necessary, the process is repeated.

The availability of current product master data, such as EAN, ILN, dimensions and weights of articles, as well as shipping units or colors are a prerequisite for the use of space management. The short product life cycles, in particular, means that establishing, and, above all, updating these data in the retail companies is very time consuming. The Centrale for Coorganisation (CCG) in Cologne with equal representation of institutions from the retail sector and industry has a project, *SINFOS product master data pool,* that offers support here. Industry companies transfer standardized and machine-readable article information to the SINFOS CCG data pool and so are themselves responsible for the quality and relevance of the data in this pool. Retail companies and other interested parties can obtain article information from the SINFOS data pool. They can then store and process these data in their own information system. SINFOS thus permits the standardized exchange of article master data between industry and the retail sector. This avoids the need to re-enter these data. SINFOS uses exclusively EDI messages based on the EANCOM Standard for the representation and transmission of article master data. A Web-based alternative, which, however, is not suitable for mass data input, allows companies to use SINFOS quickly and correctly even without having their own EDI know-how.

7.1.5 Efficient Promotion

Efficient Promotion (EP) uses a partnership-based cooperation between manufacturer and retailers to try to match all activities of the sales promotional agreement along the supply chain, i.e., consider the needs for both retailers and consumers (Möhlenbruch et al., 1999). Manufacturer promotion policies not coordinated with the retail sector result in high stocks of goods. Attractive promotion prices induce retail companies to buy more articles (forward buying) that they can later sell to the consumers at normal list prices or resell to smaller retailers (diverting). However, such savings of the retailers are reduced or even canceled by other increased costs (e.g., storage costs) along the supply chain. As part of the EP, the promotion practices of retailers and manufacturers are now matched with the aim of replacing the stocking of articles at promotion prices in favor of a more efficient system.

The agreements made between industry and retailers as part of the EP do not only apply to the type, object, time and duration of the promotional agreements, but also to the use of advertising, such as fliers or TV spots. Assortment assessment and cost transparency, in particular through the use of the activity-based costing, are the major factors of the EP. An expression of the successful agreement is the observed replacement of excessive – and eventually brand damaging – promotions with everyday low prices, linked with an analysis of repeat purchases and frequent-shopper programs to reward customer loyalty.

Everyday low prices have advantages for all market participants:

- Consumers are always offered a reasonable price without needing to make intensive comparisons and then buy from several retailers. This saves costs, time and increases customer satisfaction.
- Retail companies can establish a loyal customer base without entering the loss zone.
- Manufacturers can better balance the production capacities because constant prices avoid jumps in the demand.

Promotional agreement activities are closely linked with the logistical processes. For example, because special packaging and other pallet sizes are often used for promotions, agreements must be made between manufacturers and retailers to ensure the efficient handling of the promotion goods. This should avoid increased promotion sales being counteracted by increased handling costs. On the other hand, logistics strategies of ECR highly depend on EP, because it provides an instrument to avoid high stock levels and reduce the volume of promotion goods

The Collaborative Planning, Forecasting and Replenishment (CPFR) Committee of the US VICS (Voluntary Interindustry Commerce Standards; information in the Internet under http://www.vics.org/) consortium released mid-1998 a procedure to support the EP strategy. The goal of the CPFR is "*to create collaborative relationships between buyers and sellers through co-managed processes and shared information*" (CPFR 1999). This procedure describes the exchange of sales volume, forecast, procurement and promotion planning data that the business partner then can use to make an early comparison with its own expectation for the sales market.

7.1.6 Efficient Product Introduction

Efficient Product Introduction (EPI) is an ECR strategy in which industry and the retail sector together develop and introduce new products with the aim of reducing costs, time and risk. Industry often develops new products isolated from the retail sector, even though its intensive contact with the customer means that it has a wealth of knowledge and experience about the actual requirement for new products. The stagnating sales and the increasing competition pressure forces many manufacturers to continually develop new products and introduce these in the market, even when such products by no means always represent innovations. On the other hand, the possibilities of placing new articles in the retail sector are reduced as result of limited sales areas and restricted organizational and personnel prerequisites. Thus, the purchasers in the retail sector have more selection capabilities and consequently more power in negotiations with the manufacturers. Consequently, the success of the product innovations also depends significantly on the acceptance by the retail sector. Thus, a close cooperation of the industry and the retail sector is desirable. Originally, EPI focussed only on the introduction of new products. However, the previously described conditions mean that an extension of the consideration to the cooperative product development is logical. The fast response to consumer acceptance or rejection should reduce the flop rate for the introduction of new products. A special goal of EPI is the cooperation of retail companies with manufacturers for the development and introduction of dealer's brands.

The agreements for pricing, optimum package and pallet sizes, and estimates of possible sales are central points of the cooperation in EPI. A close binding also exists to the other marketing strategies of the ECR, in particular to EP, for example, not only for advertising, determining of test markets or introductory prices, but also for the space management. The process of the product introduction can be divided into four phases. Initially an *assessment of the product idea* and a *general planning of the marketing concept.* The *product packing is determined* parallel to the development of the marketing concept. The *realization of the marketing concept* and *the listing by the retail sector* then follow. Because data important for the company success, such as product development concepts and customer structures, must be made public, there is conflict potential in the cooperation for the joint development of products. Both parties must exhibit not only a high degree of mutual trust and openness, but also some readiness to take risks.

7.1.7 Summarizing Assessment

ECR designates a bundle of various strategies. Some of the individual methods have been a fixed part of the business discussion for quite some time. These are independent concepts that have been used successfully in practice. ECR also does not provide any fundamentally new tools to support the strategies. Scanning, electronic data exchange, space management and cross-docking are technologies and procedures already established in practice that realize significant saving potentials even when used in isolation. The true capability of the ECR concept lies in the integration of individual strategies used in isolation and in the consistent orientation on customer benefits. Significant business success can also be achieved through the use of just the logistics or marketing strategies. However, such an isolated use must always remain suboptimum. In particular, the example of the previously described marketing strategies of the ECR makes clear that

- significant interdependencies exist between the individual strategies, e.g., EPI is not sensible without EP,

- the organizing of a logistics systems must also always be considered in conjunction with the defined assortments,
- the efficient promotional agreement is also not possible without integrating the logistics.

The consistent orientation on customer benefits associated with ECR forms an integration frame – a common "central theme" – for all substrategies and so benefits from synergy potentials resulting from the bundled use of the strategies.

An important consideration for the evaluation of the ECR is the question of the suitability of this concept for certain assortments. ECR is well-suited for the groceries area. This normally contains mass products characterized with continuous demand. ECR appears to be particularly advantageous for perishables. The season factor is also a significant consideration for the evaluation of the product suitability for an ECR concept. Products with exclusively seasonal demand can scarcely be produced in synchronization with the demand, because this would almost certainly produce delivery bottlenecks. Consequently, the strategies for such merchandise categories must be modified as shown by the quick-response concept discussed in Section 7.1.2. Irrespective of their product properties, the achieving of a critical mass, namely an adequately large sales volume, is essential for all products. Many ECR procedures, such as cross-docking, require specific quantities in order to achieve increases in efficiency and savings.

Compared with the USA and Great Britain, Germany trails in the ECR development. To some extent this can be attributed to a certain disenchantment after the first pilot projects. Indeed, the experience gained and the potentials realized in the USA cannot be passed directly to the Germany economy. On the one hand, this can be attributed to the different structures of the German and American retail companies. On the other hand, the success of ECR pioneers, such as Wal-Mart, is often attributed just to its management concept and it is forgotten that other factors, such as location policies, employee participation and willingness to make innovations, have contributed to the success of such companies. In addition, legal restrictions in Germany limit a significantly stronger customer orientation, for example by providing customer discounts based on the sales volume per year, quarter or month.

7.2 Realization of Efficient Consumer Response with SAP Retail

7.2.1 Standard Functionality for the Support of Cross-Company Logistical Strategies

As mentioned previously, standardized *identification systems* are a prerequisite for a successful implementation of ECR. SAP Retail permits article numbers with up to 18 digits, which allows the generating of in-house numbering systems. In addition, all types of the international article numbering schemes (EAN) are always supported in the article administration, such as manufacturer EAN, in-store EAN, short EAN or perishables EAN (also refer to “EAN/UPC” in Section 3.2.5). Because different packing sizes (pallet, packaging carton, unit) can arise in various logistical processes, for example, goods receipt and sales, SAP Retail allows the assignment of EANs to packaging and quantity units. This supports scanning at all logistical levels – an important prerequisite for an efficient handling of the goods flows along the supply chain.

A further important prerequisite for an efficient inter-company cooperation is the support of *electronic data exchange* (*EDI*) using standard software. SAP Retail permits this communication using the integrated EDI communications interface (also refer to Section 4.6). Thus, purchase orders, order confirmations, shipping notifications, invoices, etc., can be sent and received automatically. Similarly, product information from manufacturers for updating the article master data, for example, can be transferred from the above mentioned SINFOS data pool into SAP Retail.

It is possible to define in SAP Retail which EDI messages can be exchanged with which partners. EDI subsystems perform the appropriate conversion of EDI messages to a standard relevant for the retail sector (EDIFACT, ANSI X.12 or EANCOM), the administration of partner profiles, the exchange of messages with other systems and the archiving of EDI messages.

The central prerequisite for the functioning of a pull-oriented logistics system is the acquisition of the events that change the inventory accurate at the article, quantity, time and store level. Consequently, the connection of the DP systems at the POS to the central DP system is a decisive factor for ECR. The *integration of POS and store retail information systems* in SAP Retail has already been discussed in Section 4.7.

The *inventory management* is an elementary part of the ECR process. It is primarily responsible for supporting the store replenishment. SAP Retail supports two procedures for the inventory management: exact article inventory management and inventory management at the sales price level (also refer to Section 3.4.2).

The exact article inventory management is of particular interest for the support of ECR logistical strategies. This inventory management can be based either on quantity or value. Because SAP Retail always maintains stock levels at the storage location level, the stocks of goods in the individual stores can be managed without difficulty. SAP Retail distinguishes between different stock types, such as freely-usable stocks, return stocks or consignment stocks. The inventory management process is initiated after every action that changes the stock level, for example, after the POS upload. Stock data are always maintained up-to-date.

SAP Retail provides various means to realize *logistical concepts* for the Efficient Replenishment. The following section describes some important options. The push and the pull principle (also refer to Section 7.1.1), which both can be reconciled with the aims of the ECR, can always be used for the goods distribution in retail companies with stores. The push principle is used primarily for the first distribution of new articles, for the distribution of promotion goods and remaining stocks and for the allocation of non-replaceable goods subject to fashion trends. SAP Retail permits distribution using a allocation table to initiate the planning and the procurement actions. It is possible to specify allocation rules with recurring shares (related to articles or merchandise categories) defined for the stores. The use of the allocation table allows the generation of procurement documents (vendor orders) and issue documents (warehouse orders and deliveries) as follow-on documents of the allocation tables. Because a procurement already exists when the allocation table is created with reference to a purchase order or a shipping notification, issue documents are generated only as follow-on documents. During the generation of the follow-on documents, the distribution data are updated in the system. These data are used after the goods receipt to control the goods flow in the distribution center. During the goods receipt, the distribution data are initially adapted automatically or interactively to the quantities actually delivered. The goods are then distributed to the customer. SAP Retail provides several processing procedures for this customization. For example, the goods are either stored or moved directly to the goods issue using cross-docking or flow-through .

The following business transactions (item types) can be realized using allocation tables (Figure 7-3):

- **Allocation as third-party business:** The distribution is planned using an allocation table. The goods are then ordered with an external or internal vendor. Because the vendor supplies the customer directly, no warehouse orders are generated for the customer in this case.

- **Two-stage procurement (goods flow from the vendor to the customer via the distribution center):** In this case, the stores' purchase order with the distribution center also initiate a purchase order from the distribution center with the external vendor.

- **Goods flow from the distribution center to the customer (stock reduction):** In this case, the procurement of the goods has already been completed. An allocation table is used to plan the distribution of the available warehouse stock. Warehouse orders are then generated for the customers.

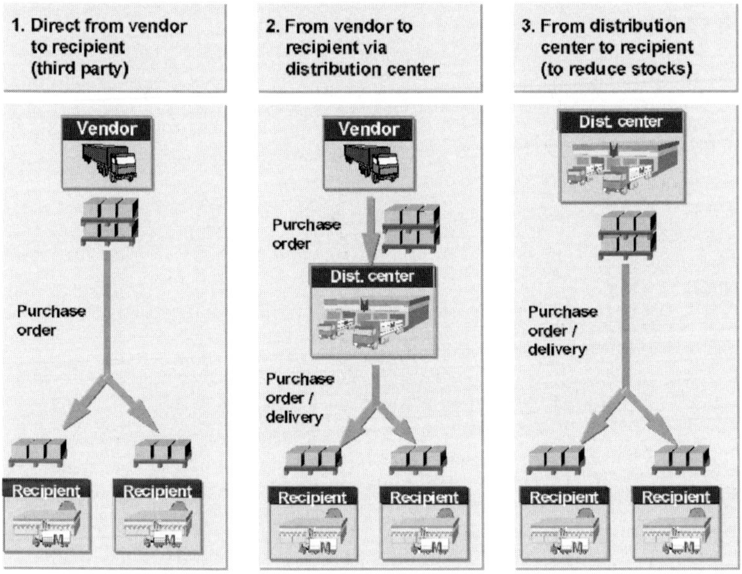

Figure 7-3: Possible business transactions in the allocation table

The allocation table can also be used to handle planned returns (e.g., recall actions). In this case, return orders (vendor returns or store returns) are generated as follow-on documents.

It is also possible to inform the stores about the planned quantities (notification, allocation table notification). The stores can modify the details (request notification) which also results in partial support of the pull principles. In particular, the combination of returns processing and request notification has a relevance in practice that should not be underestimated: the head office specifies the article (e.g., for a recall action), to which the stores add the quantities.

For the determination of the required quantities (pull principle), issue documents (sales orders or warehouse orders) already exist for the stores. *Collective purchase orders* are produced as procurement documents by accumulating the issue document quantities per article and distribution center. The distribution data evaluated during the goods receipt as described above are also updated in the system during the generation of the collective purchase orders. Figure 7-4 outlines the standard procedures for the distribution.

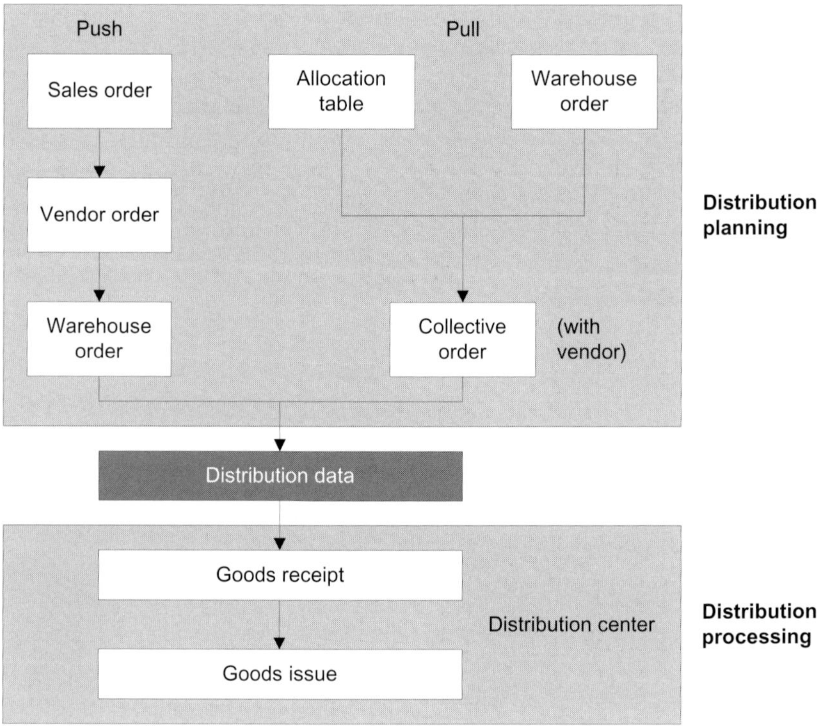

Figure 7-4: Push and pull principle for the goods distribution

The goods procurement to match the demand is a main requirement placed on the *purchasing* (this has already been discussed in detail in Section 3.3). Sales recording at the article level and up-to-date inventory management guarantee well-founded base data for this function. Major purchasing tasks are the stock monitoring and the automatic determination of order proposals. The *reorder point planning* supported by SAP Retail provides a permanent comparison of the current stock level with a reorder point. If this level is fallen below, an order proposal will be generated, unless this has already been done. The reorder point is determined from a safety stock and the expected sales during the replenishment time. Reorder point and safety stock are central control quantities for this purchasing procedure. SAP Retail allows them to be modified automatically depending on the appropriate sales and delivery situation. For the *stochastic purchasing*, the forecast values also form the basis for determining the sales order quantities. The forecast is based on historic values that SAP Retail uses to automatically offer the purchaser an appropriate forecast model (e.g., constant, trend, season or trend-season model).

As part of an efficient ER realization, it is important to keep the time interval between the generation of the order requests and the production of the sales orders short. SAP Retail allows purchase requisitions to be converted automatically into sales orders, provided an appropriate indicator has been selected in the article and vendor master data. SAP Retail also provides the capability to initially generate order proposals from purchase requisitions and allow these to be processed interactively by purchasers. Because this procedure is time-consuming, it has more secondary importance in the context of an ECR implementation.

The ECR-conform *replenishment procedure*, as method for the demand-oriented merchandise supply for customers, is primarily based on the stochastical purchasing. All goods movements are considered here. In particular, the sales recorded at the POS in the store are transferred with EDI to SAP Retail, processed in the POS interface input (also refer to Section 4.7) and posted as goods issues. A replenishment planning that determines the articles relevant for replenishment is performed at regular intervals. These are all articles for which that actual store stock has fallen below a defined (static or dynamic) target stock. For these articles, the replenishment requirements are calculated using the stock and sales data. The static target stock is specified as parameter in the master data; it specifies how large the number of goods on stock in the store should be after the goods receipt. The use of the static target stock is possible, for example, when sales are constant. The dynamic target stock requires a forecast of the sales prior to the replenishment planning. It is formed from the sum of the forecast sales within the target range of coverage (time duration between the goods receipt for the current planning and the next goods receipt) plus a safety stock. This procedure is particularly suitable for a non-constant sales behavior.

The use of this method is not restricted to the retail sector. Indeed, this procedure can also be used by wholesalers and brand-name article manufacturers. Thus, SAP Retail supports the handling of the procurement logistics based on ER principles, in particular, the *Vendor Managed Inventory (VMI)* in which the vendor undertakes the purchasing of its articles on behalf of the customer. In this case, the customer, namely the retail company, sends the current stock and sales data for an article to the vendor. The required period-related sales data come from an information structure of the Retail Information System (RIS) described in Chapter 5. It is also possible to transfer information relating to open purchase orders that SAP Retail provides for the vendor's associated article. EDI is used to transmit the data to the vendors. The manufacturer's SAP system or other EDI-conform systems process the received data. In the case of SAP, the sales data are accepted in the information structures of the information systems. For example, they are available here for forecasts or statistical analyses. The stock data are updated in the article data to permit the replenishment planning for the retail company. The replenishment needs are initially determined during the planning. Firstly, the stock quantity can filled up to a fixed (static) target stock, and, secondly, the target stock can be maintained variable depending on the sales forecasts (dynamic target stock); this latter procedure should be favored with regard to the optimization of the supply chain. In this case, the manufacturer needs the sales data from the retail partners. Once the vendor has created a sales order, it sends a purchase order confirmation by EDI to the retail company. A purchase order for the received purchase order confirmation is now generated in the retail partner's SAP Retail system (Figure 7-5).

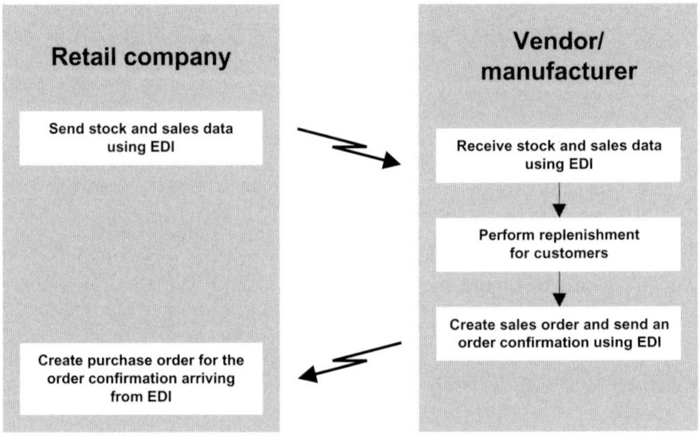

Figure 7-5: VMI via EDI

The use of transit terminals or goods distribution centers is one of the central concepts of the ECR. SAP Retail supports two procedures for the goods processing in a distribution center. For *cross-docking*, the goods have already been picked by the manufacturer and are not repacked in the distribution center between the goods receipt and goods issue (also refer to "efficient replenishment" in Section 7.1.1). In the second method, the *flow-through procedure*, the delivered goods are not prepicked. After goods are received, they are transported to a repacking zone, repacked there appropriately for the stores (picked), and then transported to the goods issue area. Goods are not stored in this procedure, too. SAP Retail provides functions for the planning of the picking activities that allow the expected picking load to be displayed per shipping point. The data needed here, for example, the processing time per quantity unit, can be specified in the article master data. The handling of the picking is supported by the creation of picking lists. An appropriate indicator in the allocation table item controls whether or not cross-docking is to be used. This indicator is passed to the purchase order. If the distribution is performed exclusively using cross-docking, SAP Retail can avoid the integration of a warehouse management system. Delivered quantities are then distributed to goods recipients automatically or interactively.

The *goods receipt planning* permits the avoidance of bottlenecks, for example, at the goods receipt ramps, and thus is a central element in the optimization of a supply chain. The goods receipts planning in SAP Retail is based on the purchase orders scheduled for the expected delivery date.

SAP Retail supports the *entry of the goods receipts* both with and without the associated preliminary documents (e.g., purchase orders). The goods receipts can be entered in SAP Retail with reference to a purchase order. Here, the data from the purchase order are passed as default values for the goods receipt. This simplifies the control of the goods receipt, for example, for overdeliveries and underdeliveries. Search functions simplify the assignment of preceding documents to the goods receipt (the document search can be made, for example, using the article short text or the EAN, the vendor number or the delivery date). To rationalize the goods receipt, scanning can be combined with mobile data entry systems. This greatly simplifies the entry of the goods receipt. In general, scanning and mobile data entry systems can be used extensively in the merchandise logistics, for example, for the picking or the goods issue. SAP Retail provides interfaces to such systems.

Several alternatives are provided to support the *goods receipts in the store*. Firstly, the goods receipt can be made as a third-party delivery. In this case, the changed store stocks are updated in the central SAP Retail system using the store goods receipt message. Although third-party deliveries are also processed in conjunction with ECR, they produce a lower loading of the transit terminals. However, the positive cost effects for goods distribution centers can be attained only for certain turnover quantities of goods. SAP Retail also supports the goods receipt of warehouse articles in two ways (articles passing through the transit-terminal can also be considered as warehouse articles). As part of the two-step stock transfer, the store delivery is initially posted as goods issue in the distribution center; the posting of the goods receipt in the store is made only when the store goods receipt message arrives. For the single-step stock transfer, the posting of the store goods receipt is performed parallel to the goods issue in the distribution center. As previously for the entry of the goods receipt in the transit-terminals, the reference to preceding documents can also be established for the store goods receipt.

7.2.2 Shelf Optimization Interface

In particular to support the Efficient Assortment, SAP Retail provides an interface to integrate special programs for shelf optimization. A large number of software packages are available for a computer-supported shelf optimization (space management). Such packages determine and manage the optimum placement of articles in stores. Here, these programs calculate the ideal placement width of individual articles within a shelf using the optimization criteria previously described in Section 7.1.4. Although SAP Retail does not provide any module of its own for shelf optimization, the most important programs, such as Apollo, Intercept or Spaceman, can be integrated to provide a data flow in both directions.

The *layout modules* defined in SAP Retail provide an important foundation for the integration. A layout module can represent one or more sales shelves. Thus, layout modules group goods that are to be presented together. A placement sequence can be specified in the layout module. SAP Retail provides two types of layout modules:

- *General layout modules* are used for multiple stores. This allows the uniform presentation of the goods in the stores for distribution chains.
- *Store-specific layout modules* apply only to one store.

The physical assignment of the layout modules defines the layout for a store or department. A single layout module can appear more than once in a store's layout.

Data can be exported from SAP Retail into the external Space Management program with either direct program call using BAPIs that have read-only access to the SAP data files (read BAPIs) or with OLE using Microsoft EXCEL. Important article data that are transferred from SAP Retail to the external system include article numbers, article short text, article sizes, quantity units, prices, sales data (by volume and value).

The shelf optimization program analyzes the shelves within the previously selected layout module using key figures and then performs optimizations. A layout module in SAP Retail corresponds to a planogramm of the shelf optimization. Finally, the optimized data – such as facing (linear shelf space quantity per article), safety stocks and maximum stocks – are returned to SAP Retail by calling the posting BAPIs to update the SAP data file. This interface provides various expansion levels. These expansion levels range from the simple information about the shelf stock and the analysis of the shelf loading and profitability, and finish with the new placement or re-placement of articles and an interactive shelf updating.

To summarize, a *shelf optimization* process can be represented as follows:

1. Identify the reasons for the shelf optimization.
2. Select the data of a layout module to be transferred to the shelf optimization program.
3. Perform the external shelf optimization.
4. Report back the result data.
5. The data are accepted, discarded or reprocessed interactively in SAP Retail.
6. If required, newly selected articles are listed for the store. Because the layout modules can also apply to several stores, different listings are possible.

7.2.3 Promotion Handling

The Retail component for promotions is specially tailored to Efficient Promotion. It permits the planning, execution and monitoring of temporary price reductions for the sale of goods that have been purchased on favorable terms, to reduce the excess stocks or as advertising for the store. Promotions are planned at company level and the data are then downloaded to the various stores at the appropriate time.

The *promotion handling* covers the following four steps:

1. Promotions are created in the system with general data, such as promotion type, planned key figures, product catalogs, theme, areas and dates when applicable. The articles associated with the promotion are added as promotion items. To simplify the promotion planning and analysis, promotions can be assigned to specific promotion themes. Assortment structures or marketing considerations can be used to determine the themes, for example, "household goods", "everything for baby", etc.
2. Important planning data, such as number of articles, the promotion sales volume, total quantities or logistical units, can be specified in the promotion planning. References can be made here to earlier similar promotions. The product catalog planning supports the administration of product catalogs. Special product catalogs can be created in SAP Retail for promotions. Sales aids, such as special labels or tags, can be assigned to the promotion articles.
3. To support running the promotion, purchase and sales prices, and the listing of the promotion goods and allocation tables are activated. A promotion announcement is sent to the participating stores. Finally, SAP Retail realizes the separate administration of the promotion stocks.
4. The analysis of the promotions is based on the information structures of the RIS. A summary can be made, for example, according to stores, merchandise categories, articles, promotions or promotion themes. Relevant key figures for the analyses are, for example, the promotion sales, the total of the goods receipts per vendor or the purchase order volumes. As already mentioned for the handling of the RIS, flexible analyses can also be generated.

The following summary characterizes the functions provided by SAP Retail to support promotions:

- automatic download of promotion prices into the POS systems at a specified date of validity and automatic return to normal prices when the promotion has ended
- possibility to control the degree of involvement of stores (obligatory goods quantities per store, optional quantities, optional involvement)
- assignment of special contracting or sales agreements for promotions
- management of special price offerings such as "three articles for the price of two"

- possibility to specify different validity periods and/or delivery dates for the stores participating in the promotion
- automatic allocation of goods to stores
- a separate inventory valuation of promotion and normal stock
- automatic updating of sales quantities and other promotion data while a promotion is running
- availability of various reports and analyses to evaluate promotions.

7.2.4 Support for the Introduction of New Products

The cooperative introduction of new products as part of the Efficient Product Introduction (EPI) ECR strategy often has a project character. Consequently, project management systems, such as the integrated *project system* (PS) from SAP R/3, appear suitable to support this strategy. We can only provide an outline of basic functionalities of this module. The PS supports all phases of a project. Central structures in this system are network diagrams and project structure plans. The project structure plan represents the services to be provided by the project as hierarchy and so forms the basis for the organization and coordination in the project. Network diagrams are used to schedule workers, capacities, materials, equipment and services for the project. The PS guarantees a real time project tracking starting with the project structuring, including the procedure planning, and ending with the controlling. It independently monitors the availability of the finances, capacities and materials needed to perform the project.

PS supports the central requirements placed on project management systems, such as the definition of milestones, the management of specifications, the monitoring of project costs and the assignment of budgets. It also provides an information system that allows important reports and analyses about projects. Dräger (1998) contains a detailed description of the R/3 project system.

An important function for EPI is support for the *listing* of newly introduced products. SAP Retail provides the following capabilities to accept articles in the assortment:

1. by direct activation from the article updating (entered articles are not automatically listed)
2. with integrated functions for the interactive updating of article assortments
3. by changing parameters in assortments that affect the assortment form
4. with allocation or promotions (implicit listing).

A subsequent listing is also possible, for example, when a manufacturer delivers a new article instead of an old article that is no longer produced. The site can possibly accept the new article, even when this is not currently contained in the approved listing.

7.2.5 mySAP.com – additional Solutions for e-business

Recently SAP has divided its product spectrum in 9 categories including additional software components for specific areas of use besides the well known R/3 modules. This product spectrum is called mySAP.com

- The products of the category mySAP E-Business Solutions are of special importance for the realization of ECR and Category-Management-Solutions. This category includes besides the R/3 application modules the following regions:mySAP Business Intelligence (refer to chapter 5)
- mySAP E-Commerce (refer to chapter 6, the component SAP B2B Procurement is to be presented briefly in the following)
- mySAP Product Life-Cycle Management
- mySAP Supply Chain Management
- mySAP Customer Relationship Management

Important components of the both last areas are to be presented briefly as follows. The mySAP Supply Chain Management is related with the well known logistics modules of R/3 (MM, PP, SD...), summarized by the designation logistics, the new products Advanced Planner and Optimizer (APO) and the Logistics Execution System (LES). These products are to be presented in the following section.

7.2.5.1 mySAP E-Commerce

SAP – Business to Business Procurement (SAP B2B Procurement)

The SAP product B2B Procurement allows authorized employees to use an intuitive interface to query both goods and services for the areas of maintenance, repair and company consumption. It is used over the Internet. Online catalogs from various providers can be integrated. Section 6.6.2 discussed these components in more detail under the aspect of the electronic commerce. Figure 7-6 shows the user interface of the B2B Procurement with display of the shopping basket.

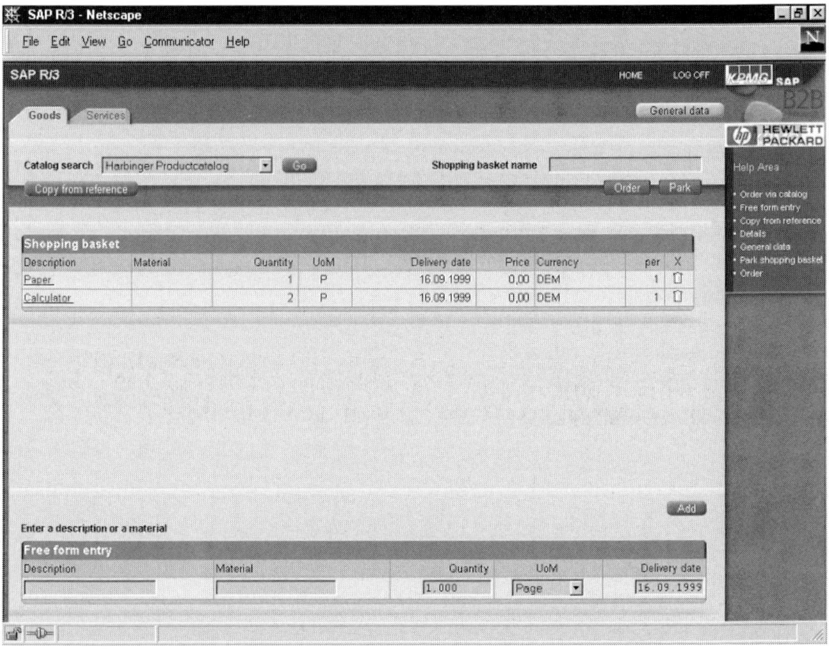

Figure 7-6: SAP Business to Business Procurement

7.2.5.2 mySAP Supply Chain Management

mySAP Supply Chain Management groups products that are used to monitor and optimize elements of the supply chain. This provides a framework for structured data storage and data analysis, planning, forecasting, simulation and optimization, control of the operative procedures and for an efficient customer support and sales. The following products provide functions that are also relevant for retailing.

SAP - Advanced Planner and Optimizer (SAP APO)

The SAP Advanced Planner and Optimizer with a user-friendly interface and the direct integration in the SAP Retail system or in third-party products provides an efficient modeling, planning and monitoring of the supply chain processes with various levels of detail. The architecture shown in Figure 7-7 can be used to derive the functionality and uses of the SAP APO.

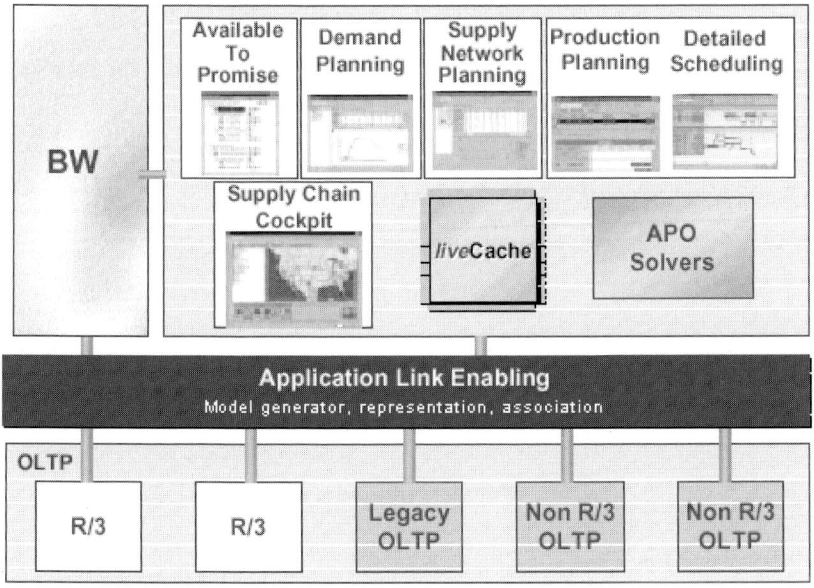

Figure 7-7: SAP APO architecture

The SAP APO currently contains the following components:

- *Supply Chain Cockpit:* a graphical "instrument board" to model, represent and control the supply chain
- *Demand planning:* a grouping of statistical forecasting techniques and demand planning functions to create precise forecasts
- *Supply Network Planning and Deployment:* a planning method to create tactical plans and ordering decisions that cater for the complete logistical network
- *Production Planning/Detailed Scheduling:* scheduling for the subsequent sequence planning and final assignment of production resources to create optimum production schedules
- *Available To Promise):* multi-level availability test for material and capacities (including transportation capacities)

The *Supply Chain Cockpit* is the graphical user interface of the APO. This provides decision makers with real-time information about the current status of the supply chain and control capabilities. The Supply Chain Engineer integrated in the APO can be used to create graphical displays of the logistical network by selecting and connecting various types of network nodes (vendor locations, distribution centers, stores, itineraries, etc.). Data objects (e.g., resource capacities, goods inventory, vendor schedules) that contain information from internal and external sources are assigned to the network nodes. The view of the supply chain within the company can be detailed down to the lowest organizational level. The user interface can be customized for the specific requirements of various users. Figure 7-8 shows an example of the interface. The Supply Chain Cockpit provides an Alert Monitor that uses defined exception conditions to monitor the processes along the supply chain and report any exception situations.

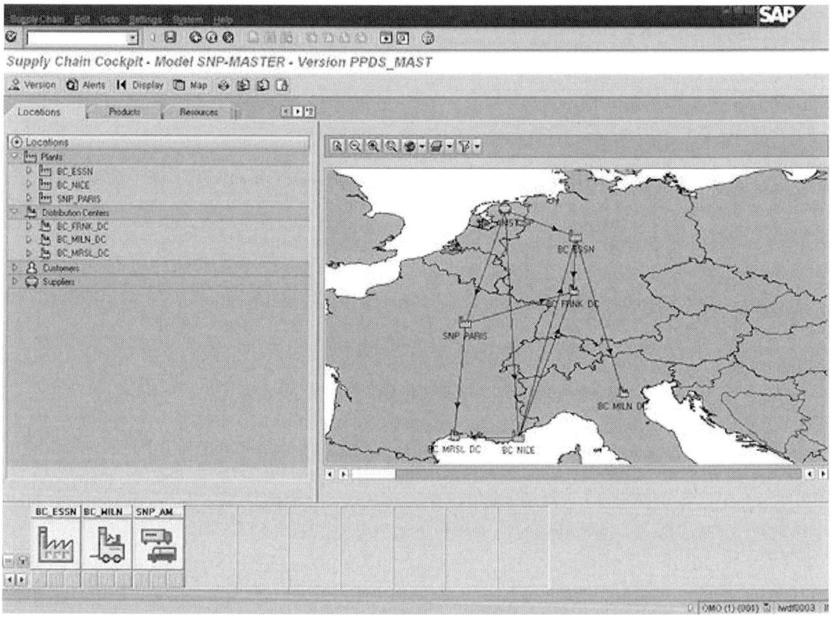

Figure 7-8: SAP Supply Chain Cockpit

The *Demand Planning* component supplies various forecasting methods and planning tools that can be used cooperatively by manufacturers and retail companies to provide early recognition of any changes in demand by consumers and trends, and to include these in the planning. In addition to creating forecasts, a life cycle management is used to support the planning for the introduction of new products and thus the EPI strategy of the ECR. Multi-dimensional data models, together with effective storage procedures and search functions, allow information to be displayed and analyzed from various perspectives and at different aggregation levels. The function for planning special offers can be used to compensate for unexpected demand fluctuations. It is also possible to use demand histories and a number of calculation models to produce forecasts for new products and to pursue cause-research for demand fluctuations resulting from changed environmental conditions.

Once the Supply Chain Engineer has been used to model the logistical network, the *Supply Network Planning and Deployment (SNP)* can be used to match supply and demand. A wide range of secondary conditions can be incorporated in the logistical network. This component provides the following functions for the efficient planning of the supply chain:

- planning and optimization in the various levels taking account of various parameters, such as transportation, manufacturing and storage capacity, works calendar and costs
- simulations with analyses, e.g., what-if analyses
- matching of demand and supply with product substitutions and selection of alternative locations
- integration of vendor managed inventory (VMI) in the logistical network
- short-term inventory and distribution optimization, e.g., for short-term demand fluctuations
- determination of stock movement recommendations based on allocations
- graphical display and control of the network nodes and inventory
- information about problems in the supply chain using an alert monitor

Whereas the Supply Network Planning and Deployment is responsible for the flow of material and resources, the *Production Planning/Detailed Scheduling* performs the planning task for the individual nodes of the logistical network. Various calculation and planning methods are provided for quantities, time and capacity planning. To improve the clarity and efficient use, various graphical user interfaces, such as an alert monitor or interactive Gantt diagrams, are available for the planning.

The *global availability check* (Global ATP) uses a rule-based strategy to match supply and demand. It provides the following three methods here:

- product substitution (rule-based selection of substitute products),
- selection of alternative storage locations, and
- assignment of scarce products and components to specific orders, markets or customers.

Further information about SAP APO can be found in the Internet under http://www.sap-ag.de/scm/apo/index.htm.

SAP - Logistics Execution System (SAP LES)

A close matching of warehouse management and transportation management, together with detailed information about goods in transport, is essential to optimize the delivery processes. The SAP Logistics Execution System supports the execution level of the logistics chains; it improves and extends the functionality of the supply chain support systems with the following components:

- LES Warehouse Management System that is tailored to the business processes for distribution centers and stores
- LES Transportation Management System that was developed for the planning and monitoring of the goods transportation (Figure 7-9).

Further information about SAP-LES can be found in the Internet under http://www.sap-ag.de/solutions/scm/les/les_over.htm.

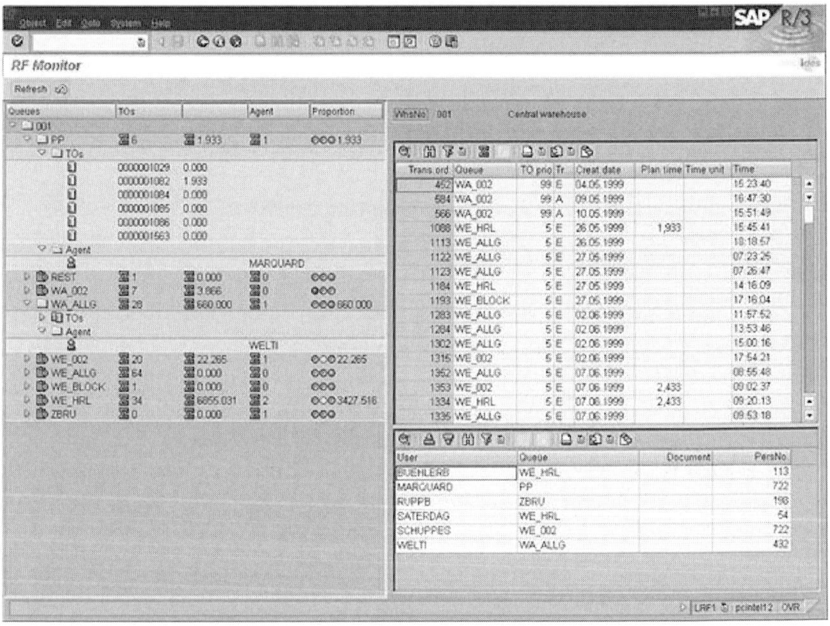

Figure 7-9: SAP LES Transportation Management System

7.2.5.3 SAP Customer Relationship Management (CRM)

Products of the Customer Relationship Management provide support to satisfy the increased demands made by customers and to increase the effectiveness and efficiency of the planning and execution of customer contacts. The products are designed to support the maintenance of customer relationships over their complete duration, starting with the market analysis and planning, including the customer acquisition and order processing, through to the customer support and service. The goal is to supply the data needed for the correct and reliable forecast, planning and satisfaction of the customer's expectations in each of these phases. Figure 7.10

shows the functionality of the SAP Service, SAP Marketing and SAP Sales products offered here.

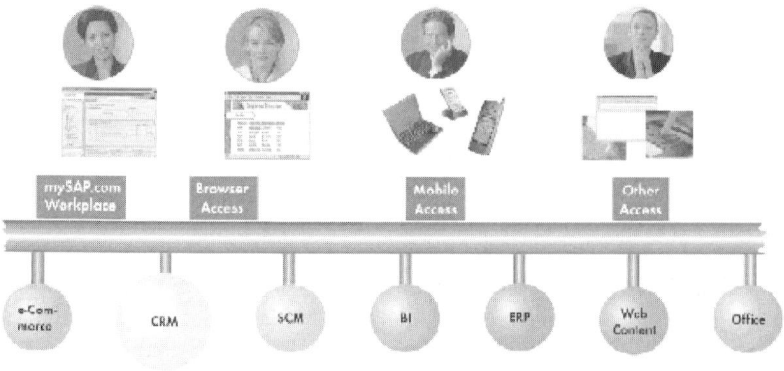

Figure 7-10: SAP CRM component within the mySAP.com component landscape

SAP's CRM-Solutions link back-office- with front-office-functions within an enterprise as well as across enterprise borders . CRM with mySAP.com consists of the following elements (refer to figure 7-11):

- A central CRM-System supporting the processing of CRM relevant business objects (e.g. business partners, activities, products).
- CRM components, such as Internet Sales, Mobile Sales, Customer Interaction Center etc., providing access to the CRM-System.
- A technological infrastructure enabling communication with other systems.

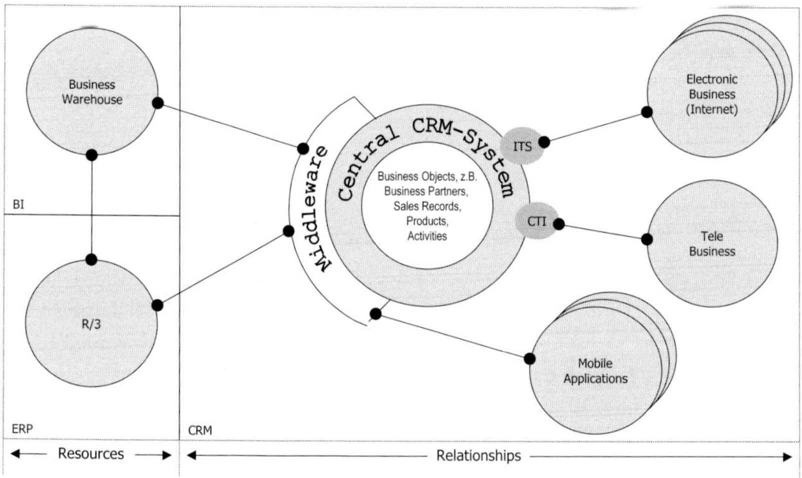

Abbildung 7-11: SAP-CRM technology landscape

The access to the central CRM-System takes place in the Internet via the Internet Transaction Server, in the Contact Center via a CTI- or Email-interface and in external duty via the CRM-Middleware. CRM-Middleware is also a component of the central CRM-System which regulates not only the data-exchange with mobile clients but also with other systems, such as back-office-, data warehouse- and Supply-Chain-Management-systems

Customer Relationship Management Workbench

The CRM Mobile Clients have been designed with a multi-layered architecture. This ensures that they are flexible and can be customized to meet the specific needs. The CRM Workbench provides the tools that are used to customize and/or develop each layer of the mobile component client software. For example, adding a new information field, modifying the layout of an existing screen, or amending the rules associated with the processing of information. The CRM Workbench provides you with a complete customizing and development environment to customize, modify and enhance the Mobile Client to fulfill your own particular needs. Figure 7-12 shows the CRM Workbench.

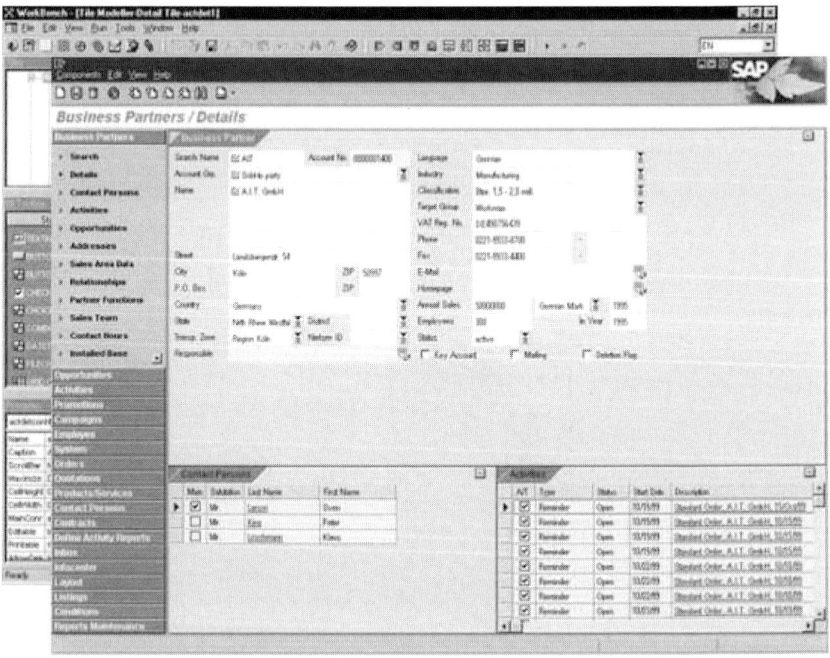

Figure 7-12: The CRM Workbench

SAP Internet Sales

Within mySAP.com, SAP Internet Sales is the e-commerce component supporting the selling side of the equation. At the same time, it is an integral part of mySAP.com. Customer Relationship Management (CRM) that enables companies to manage customer relationships effectively throughout the life cycle. SAP Internet Sales is the ideal choice for companies who have made a strategic decision to establish an Internet sales channel -one that is integrated with other customer touch points and is incorporated into the overall IT environment. The solution typically runs on a distributed software platform to spread the processing load and prevent bottlenecks.

Customer Interaction Center

The Customer Interaction Center (CIC) integrated in SAP CRM provides comprehensive call center functions. These allow call center employees to process incoming and outgoing calls using Computer Telephony Integration (CTI), and initiate the resulting business processes directly in the Retail System.

7.3 Category Management as Business Concept

7.3.1 Terms and Goal

Although the discussion about category management (CM) began already at the end of the 1980s in the USA, there is still no standard definition basis that describes this management concept. A common characteristic of many CM definitions is customer orientation by basing the merchandise categories on the demand of the customers. The cooperation between manufacturers and retail companies is a further characteristic. Finally, the process orientation represents a third fundamental property of the CM. This makes itself apparent through the avoidance of the traditional functional work assignment and the bundling of all functions for contracting, logistics and sales in a process-oriented, both company-internal and cross-company, responsibility for merchandise categories that accompanies the complete value chain (Möhlenbruch 1997).

The term *category* is understood to be a distinct assortment unit. This is accepted by the consumer as being a connected group of goods to satisfy specific requirements that is designated as merchandise category in the following section. *Category management* designates a permanent process to be realized in cooperation between retail companies and manufacturers in which the Category Manager of the retail sector manages merchandise categories as strategic business units or profit centers. The aim here is to improve results by providing the optimum satisfaction for the customer requirements. Merchandise categories in the sense of the CM are relatively independent planning and control units that are monitored and controlled using key figures.

The principal goal of CM, the consistent customer-orientation of assortments, serves, in particular, to improve the customer binding and to win new customers for the retail company. The achieving of this goal should also increase both the buying frequency of customers and the average amount of the customer purchases. In addition, a consistent customer orientation improves the image compared with competitors, and increases revenue and profits. CM should improve the quality of the assortment decisions. Associated goals are increased area productivity in the assortment range together with a more efficient and effective promotion planning. Unnecessary markdowns should be avoided and possible sales chances should be recognized faster.

From the viewpoint of the industry, CM partnerships are normally both personnel and cost intensive. Consequently, in particular for industrial companies, the question of the goals of such a partnership is of interest. Significant goals of the manufacturers with regard to the CM can be:

- influence on the actions at the point of sale
- to secure, and possible extend, the distribution of the company's own products and their contact section at the point of sale
- improvement of the placement quality of the company's own products
- gain current information about market potential, the buyer behavior, and possible substitution and complementary relations of the company's and third-party products.

The combination of different competencies by industry and retail companies can provide equal success potentials for both the retail sector and manufacturers. The retail sector has specific knowledge of the optimum combination and outsorting of merchandise categories. The manufacturers have detailed knowledge of merchandise categories, brands, and the purchasing and use habits of the consumers. However, cooperative CM also hides the danger that the retailer becomes dependent on the manufacturer (Treis et al. 1998).

7.3.2 Changes in the Organizational Structure

Merchandise categories are considered as strategic business units in the CM. This implies the need for extensive changes in the organizational structures of the companies that operate CM. Whereas traditionally the organizational structure is oriented on the functions in the companies, CM accomplishes a change to an object-oriented organizational structure, namely an orientation on the merchandise categories. Figure 7-13 illustrates this change. The *Category Manager* that plays a central role in the retail company is responsible for all purchase and sales-related activities relating to a merchandise category and, in particular, their success. The financial freedom is limited by a budget that is specified centrally or negotiated with other Category Managers. Because the Category Manager has cross-function decision-making competence, the object-oriented organizational structure has the advantage in the avoidance of frictional losses between the individual functions of the process chain. However, a disadvantage is the coordination effort needed between the merchandise categories, in particular for access to the same resources (e.g., sales areas) or for interactions between merchandise categories in sales.

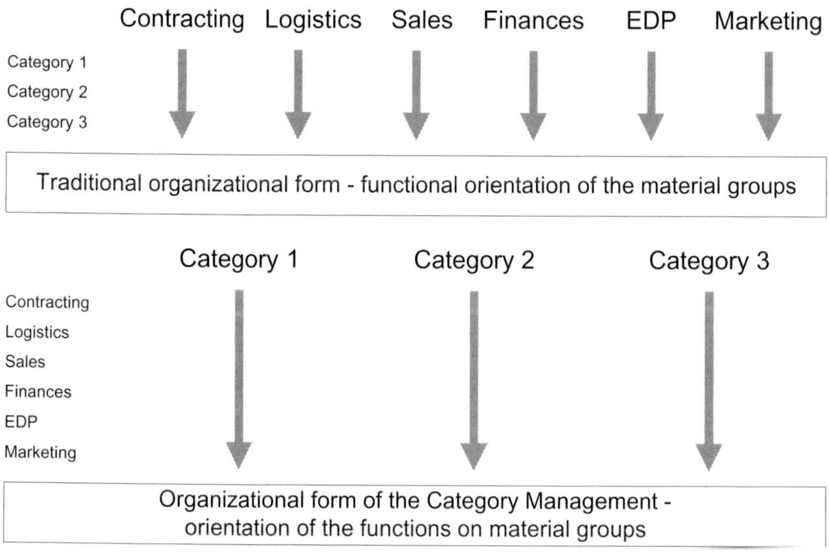

Figure 7-13: Change of the organizational forms in the retail sector by CM (Pretzel 1996)

The most important task of a Category Manager is to optimize merchandise categories and assortments and, especially, to reduce the range of articles. Here, the Category Manager uses methods from the assortment controlling based on the scanner data for the sale. The complete assortment should be significantly reduced by identifying more successful and weaker articles. Another task is the shelf optimization. Once unnecessary articles have been removed, the reallocation of the gained shelf space becomes obvious. A Category Manager is also responsible for the planning and control of promotions, and the optimization of contracting and merchandising. To allow decisions to be made on a well-founded base, comprehensive analyses must be performed in the CM. These include profitability analyses, in which merchandise categories and their articles are subjected. The Category Manager is also responsible for shopping basket analyses, in particular in conjunction with the promotion control and the placement of articles (Möhlenbruch 1997). He needs historical data for comparisons and analyzes trends and seasonal fluctuations, returns and hit lists. He must accomplish the "trick" of reconciling the product plans of the marketing with the forecasts of the sales department and the margin targets of the company management.

7.3.3 Formation of Merchandise Categories

An important prerequisite of the CM is the definition of the merchandise categories. This contains the decision which merchandise categories are to be carried in stock, how they are structured and which articles they contain. Although merchandise categories can correspond to functionally organized product groups ("spirits" or "leather goods"), they are also formed according to the criteria of the application-related association with the consumer ("everything for travel", "everything for the bathroom" or baby products). The disjunct and logically organized functional article groups tend rather to be targeted purchases to meet needs; in contrast, the article groups oriented to consumption-related association are more suitable to create demand and encourage spontaneous purchases. Needs, desires and purchasing habits of the customers should always be the central reference point for establishing and forming assortment modules. In addition to this concentration on the consumer, marketing strategies and logistical requirements must be considered here. However, a reduction of the customer orientation to their current needs would be inadequate for the development of the merchandise categories. In order to offer additional purchasing inducements, customer orientation must take account of other aspects, such as event ranges, buying behavior, buying reasons (supply or event purchase), demographic criteria, etc. (Möhlenbruch 1997).

The meaning of a merchandise category can be very different in a retail company, namely there exist various "category roles" that must be assigned to the merchandise category. The roles can be freely defined in the retail company. Standard roles are (Treis et al., 1998):

- Profile (destination): merchandise category that should form the customer's image of the retail company
- Routine: essential assortment
- Supplementary (convenience): assortment that supplements the targeted purchase for "one-stop shopping" and is profitable for the retailer
- Impulse/season: essential or supplementary assortment with temporary character

7.3.4 The Category Management Process

The CM requires the permanent monitoring, and possibly restructuring, of assortment ranges with regard to regional location factors and specific customer structures. The analysis and decision process of the CM is divided into five phases (see

Figure 7-14). As ideal cycle, it is processed iteratively. The strategic goals and general conditions are specified during the *analysis of the merchandise category*. Here, the partners from industry and the retail sector develop a common understanding of the merchandise categories and thus provide the basis for a cooperative CM. The changes of turnover figures compared to previous periods are used to derive the strength and weakness profile of the merchandise category and its future chances on the market. The role assignment is made. The *analysis of the customer potential* that follows in a second step determines the customer profile of a retail company and the purchaser profile of the associated merchandise category. In addition to an exact target group and assortment definition for each individual merchandise category, a detailed analysis of the customer wishes and the buying behavior must be performed. This allows the introduction of promising category-specific measures. The *strategy development* follows immediately. Appropriate strategies to correct existing weaknesses and to achieve the specified goals should be found. Simulation methods have attained here particular importance for the evaluation of the strategies. The planning of the promotional agreement activities and the shelf placement is also important in this phase. In the next phase, the *use of the newly developed strategy*, handling outlines are specified, implementation plans prepared and the communication to the customers organized. The implementation results are compared with the goals and modifications possibly made during the *assessment of the results* (Milde 1996, Pretzel 1996).

Figure 7-14: The analysis and decision process of the CM (based on Milde 1996)

7.3.5 Efficient Consumer Response and Category Management

CM and ECR relate to each other in very different ways. The CM often uses just the EA strategy of the ECR. However, this reduces CM to the efficient form of the assortments and their optimum placement. Nowadays, the view has been accepted that all ECR strategies can be introduced advantageously using the CM concept. Thus, CM is accepted as a "key concept for ECR" (Heydt 1998) to provide a framework for all individual strategies, as one of the organizational prerequisites of ECR, although this concept has a clear focus on the marketing strategies (Möhlenbruch 1997).

7.4 Realization of Category Management with SAP Retail

Because of the interconnection between CM and ECR, the functions and components already discussed in Section 7.2 can also be used for CM. Examples here are the shelf optimization interface or some of the New Dimension Products, such as SAP LES Warehouse Management System and SAP Marketing. The following section discusses only those additional functions and components that have not been described previously.

7.4.1 Merchandise Category Definition and Assortment Creation

Both the construct of the merchandise category and that of the assortment module in SAP Retail can be used for the CM. Merchandise category represents a possibility to structure and classify the complete assortment using company-specific factors. SAP Retail provides extensive functions in the area of the merchandise category. These include, for example, the hierarchical form of merchandise category (as true merchandise category hierarchy with the inheritance of properties to lower level merchandise categories), the use of merchandise category template and merchandise category value-only articles, and the definition of conditions at the level of the merchandise category. However, because ECR assumes that sales have been recording whenever possible for a specific article, the functionality of merchandise category value-only articles used to record sales at the point of sale that cannot be assigned to a precise article is only of subordinate interest here. Section 3.2.4 provides a detailed discussion of the merchandise categories in SAP Retail and the associated functionality.

There is a wide range of requirements for the differentiated assortment form in the retail sector. This ranges from fully uniform assortments in all areas through to unique assortments (e.g., specific assortments in each individual store). In particular in the context of CM, the requirements of the retail sector with regard to an adequate DP support for the assortment planning and maintenance have increased.

SAP Retail, with its flexible constructs for assortment creation, in particular through the extensive assortment module types (in this context, mainly the local module, the exclusion module and the rack-jobber module), provides comprehensive support (also refer to "Assortment Planning" in Section 3.5.1).

7.4.2 Category Manager Workbench

The goal for the development of the Category Manager Workbench (CMW) is to provide a user interface for the efficient support of the Category Manager in the retail company. The application developed as integral component of SAP Retail should allow the realization of a large number of the tasks arising in the CM process, such as the definition, assessment and analysis of the merchandise categories and assortments, the planning of the efficiency improvement of shelf areas or the budget planning for the categories. Figure 7-15 provides an overview of the menu structure of the CMW.

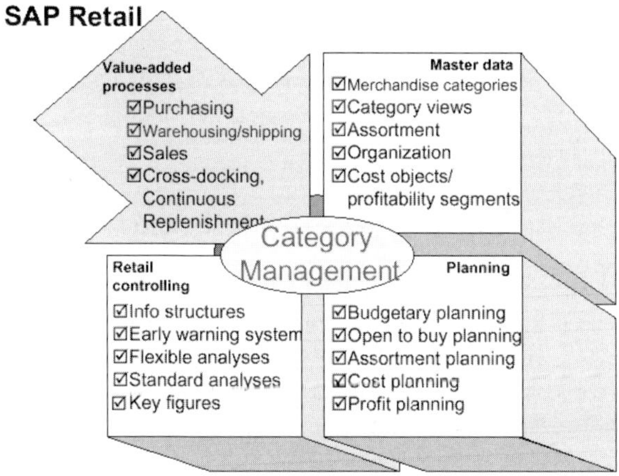

Figure 7-15: Category Manager Workbench

CM-relevant functions are integrated in a single user interface in the CMW, even though the majority of these functions are already part of the original SAP Retail. The *functionality of the CMW* can be generally divided into three areas, although the concrete form is company-specific. The updating of master data principally covers functions to define assortments, merchandise categories and their assignment to hierarchies, and for the generation of information structures in the retail information system (RIS). The previously discussed layout modules can also be used here. In addition, specific views of the CM to the master data ("category views") can be created.

CM demands, in particular, efficient instruments to support the planning and the controlling. Consequently, the planning and retail controlling function areas form the central area of the CMW. The planning is used primarily for the definition of planning figures, such as quantity sales, cost center, value sales and results plans. Because plan data are always defined on the same level as actual data, support is provided for plan-actual comparisons. The functions of the retail controlling provide the Category Manager at the operative controlling level with many reporting instruments. Planning and retail controlling use the information structures of the RIS (see Section 5.2). All functions available there, such as drill-down, pick-up techniques to generate flexible analyses, graphs, ABC analyses, etc., are also included in the CMW. The following instruments are of particular interest for the retail controlling:

- short-term retail profitability statement
- detailed season controlling for seasonal articles (e.g., fashion articles)
- shopping-basket analysis as basis for the assortment form and the optimization of the article placement
- open-to-buy analysis that allows the remaining budget for the category to be determined as difference between the planned and the actual expenditures
- exception analysis for the immediate display of critical exception situations and the periodic analysis to determine critical situations in specific time intervals
- profitability and results statement to determine the result (down to profit margin III) using the characteristics of the specified results objects.

SAP Retail provides the integrated CO-PA controlling component for the profitability report, CO-ABC for the process cost controlling and the RIS for the assortment controlling. The RIS is also used to realize the previously mentioned shopping-basket analyses, the season controlling and other retail-specific analyses.

CM designates a process that runs in close cooperation between manufacturers and retail companies. The CMW focuses exclusively on retailing companies. A version for the manufacturer of consumer goods or a version of the CMW that combines the views of the retailing sector and the manufacturers does not currently exist. This deficiency must be considered as being critical because the sole implementation of such a tool would establish an information monopoly with the retail sector which cannot be acceptable for a cooperative process. Consequently, a system that solves this deficiency and permits the joint planning is to be expected. With regard to the support of the CM process, the CMW primarily represents a solid instrument for the information procurement. In particular, the realized central functions of the planning and the retail controlling appear to be predestined here. In addition, as part of the EA strategy of the ECR, the CMW can be used to access the assortment structures.

Appendix: SAP Retail Functionality Checklist

As supplement to the main chapters of this book that described the requirements placed on an modern integrated retail information systems and mainly discussed their conceptional realization in SAP Retail, this appendix considers the functionality of SAP Retail in checklist form at the detailed level. Thus, this appendix can be used during a software selection process for an initial comparison of the SAP functionality with the individual requirements of a retailing company. Schütte/Vering/Wiese (2000) provides a practically-oriented representation of the software selection process.

The following section lists more than 300 detailed characteristics that cover not only strategic and technical aspects, but, above all, business functional areas. The retailing H-model is used as framework to structure the functional criteria.

The criteria that have been produced using experience gained from various software selection and software implementation projects represent those requirements met frequently in the retail sector (basic requirements). The criteria and the details for satisfying these are based primarily on the results of a market study for retail information systems conducted in 1999 (cf. Schütte/Vering/Wiese 2000). Further information concerning this market study can be found in the Internet under www.wws-studie.de.

It should be noted that these criteria do not describe the complete functionality of SAP Retail. Consequently, the absence of a function in this list does not necessarily mean that it is not realized in SAP Retail. Rather, the purpose of this checklist is to provide in brief form an initial comprehensive overview of the SAP Retail functionality.

1. Basic Information for the SAP Retail RIS		
Date of the first installation of SAP Retail	1994	
Number of SAP Retail installations	> 200	
System technical restrictions on the maximum installation size		
Number of users	unlimited	
Number of articles	unlimited	
Number of customers	unlimited	
Supported retailing forms	Retail sector, wholesale sector, multi-level retailing (retail/wholesale combination), mail order and agency business	
Installation according to sectors (○ = none, ⊙ = few, ● = many installations)	● groceries, perishables ⊙ building materials, lumber ⊙ metal ⊙ electrical ⊙ sanitation	● chemicals, pharmaceuticals ● mineral oil, fuels ● fashion (textiles, shoes, sport) ⊙ tires

2. DP Technology			
Supported hardware platforms for the central hardware			
☒ PC (Intel) ☒ IBM AS/400 ☒ IBM RS 6000	☒ HP-RISC ☒ DEC alpha ☒ SNI RM	☐ Sun Sparc ☒ IBM S/390	other:
Supported terminals			
☒ PC (Intel)	☒ Terminal emulation PC (Intel)	☒ ASCII terminals	other: all common terminals
Supported networks			
☒ Windows 95 ☒ Windows NT	☒ Novell ☒ Banyan	☒ IBM LAN Manager	other: all common terminals
Supported databases			
☒ DB/2 ☒ Oracle ☒ Informix	☒ MS SQL-Server ☐ Sybase ☐ MS Access	other:	

Supported programming languages			
☐ COBOL ☐ PL/1 ☐ RPG	☐ C / C++ ☐ Visual Basic ☒ 4th GL		other:

3. Service	
Software provider as contractor (hardware, software, network, implementation from a single source)	Conditional: together with partners
Hotline	Telephone, 7 days / 24 hours; also online support
Documentation	☒ printed manual, CD-ROM ☒ online manual ☒ context-sensitive online help
4. Basic Functionality	
4.1 Ergonomic Considerations	
Dialog languages of the software	☒ German ☒ English ☒ French ☒ Spanish and all other common world languages
Flexibility of the software implementation	☒ Software process models ☒ Software data model ☒ Model-supported implementation ☒ Mask generator
Flexibility of the software use	☒ Report generator ☒ Form generator ☒ User-dependent menu configuration
Adaptability of the business processes	Free configuration of the processes using predefined process modules
Authorization concepts at the level of	☒ User ☒ Organizational unit (e.g., store) ☒ Application (e.g., order management) ☒ Fields (e.g., credit limit)
Keyboard/mouse control	Complete keyboard control possible
Matchcode search	☒ Search for a field ☒ Search for several fields
Freetext search	☒ in the article master data ☒ in the vendor master data ☒ in the customer master data

Automatic document linking	Automatic linking of the procurement-side documents (order inquiry, purchase order, notification, delivery note, invoice, payment, credit) and the possible sales-side documents (quotation, order, notification, delivery note, invoice, payment, and credit)
4.2 Organizational Structures	
Multi-company capability (client capability)	Yes, unlimited number of clients
Cross-company shared (uniform within the group) master data	☒ Article master data ☒ Vendor master data ☒ Customer master data
Cross-company common (uniform within the group) hierarchies	☒ Article hierarchies (merchandise categories) ☒ Purchasing groups ☒ Sales groups
Cross-company inventory transparency	☒
Deliveries between firms in an SAP system	Manual processing, but automation of the postings possible
Representation of head office – branch structures	☒ Yes, maximum number of stores: 9,999
4.3 Articles	
Article number	Alphanumeric, default 18 digits
Article description	☒ Article short designation ☒ Article long text ☒ Multi-language article text
EAN/UPC code	☒ Manufacturer EAN ☒ In-store EAN ☒ Weight/price EAN ☒ Short EAN
Other article properties	☒ Article size ☒ Article weight ☒ Picture file assignment ☒ Sound file assignment ☒ Customer article numbers / customer article designations ☒ Vendor article numbers / vendor article designations
Article groupings	☒ Merchandise categories ☒ Vendor sub-range ☒ Sales-side sub-range

Merchandise category functionality	☒ Merchandise category reference article ☒ Merchandise category value article ☒ Cross-merchandise category value article
Merchandise category hierarchy	☒ True hierarchy with variable number of hierarchy levels (3 levels predefined) ☒ Reclassification of articles possible
Reference to substitute/alternative article	☒ Assignment time-controlled
Article variant management	☒
Structured articles	☒ (Sales) set ☒ Lot ☒ Display
Build sales sets from individual articles	☒ Price as sum of the individual prices ☒ Price as special set price ☒ Inventory management at the level of individual articles ☒ Inventory management at the set level
Management of sales aids	☒
Packing units per article	Several hierarchically linked packing units
Batch management	☒ General batch classification based on use-by date, date code and metal content as percentage ☒ Batch sorting using various characteristics, e.g., batch number, use-by date, date code ☒ Stock reservation for a specific batch
Dangerous goods classification	☒
5. Business Processes	
5.1 Contracting	
5.1.1 Vendor	
Vendor number	Alphanumeric, default setting: 10 digits
Vendor addresses/roles	☒ Order recipient ☒ Goods supplier ☒ Invoicing party ☒ Condition granter ☒ Payment recipient ☒ Bonus creditor ☒ Contact partner ☒ others, freely definable

Store-dependent assignment of different order addresses of a vendor	☒
Vendor groups	☒
Vendor hierarchies	☒
System-supported vendor assessment	☒
5.1.2 Contracting Conditions	
Contracting price scales	☒ Quantity scales ☒ Price scales ☒ Arbitrary number of scale levels
Possible levels of the contracting price definition	☒ Cross-company ☒ At company level ☒ Specific at store level
Contracting discounts	☒ Percentage ☒ Absolute ☒ Bonus quantities ☒ Early-order discounts ☒ Discount scales (value and quantity related)
Possible levels of the contracting discount definition	☒ Cross-company ☒ At company level ☒ At specific store level Definition possible with reference to a single article, a vendor sub-range or vendor assortment
Time control of contracting	☒ Time control of the prices ☒ Time control of the discounts ☒ Advance entry of future changes ☒ Conditions history
Conditional conditions (exclusion conditions)	☒
5.1.3 Subsequent Remuneration (Contracting Bonus Agreements)	
Definition of subsequent renumerations (bonuses)	☒ Time control of the bonus agreements ☒ Quantity scales, value scales ☒ Bonus exclusion
Level of the bonus definition	☒ Article ☒ Vendor sub-range ☒ Vendor assortment

Updating of bonus claims	☒ for purchase order ☒ at goods receipt ☒ at invoice verification
Settlement of bonus claims	☒ One-off and periodic settlement ☒ Intermediate settlements during the run time
5.1.4 Contracting Outline Agreement	
Outline agreement types	☒ Quantity contract ☒ Value contract ☒ Scheduling agreement
Support for overlapping or parallel outline agreements	☒
5.2 Purchasing	
Purchasing according to....	☒ Vendor ☒ Vendor sub-range ☒ Merchandise category ☒ Customer
Purchasing types	☒ Manual purchasing ☒ Automatic purchasing with manual release ☒ Automatic purchasing and initiate purchase order
Specification of the purchasing type at the level	☒ of the merchandise category ☒ of the article
Procedures for the automatic purchasing	☒ Replenishment purchasing ☒ Time of ordering purchasing ☒ with trend model ☒ with season model ☒ with trend/season model
Two-stage purchasing (collective purchase orders) / purchase order bundling	☒
Investment buying	☒
Find best price for alternative sources of supply	☒
Limit analysis at the level	☒ of the article ☒ of the merchandise category ☒ of the vendor ☒ of the MRP controller
Matrix acquisition for articles with variants (e.g., jeans)	☒

5.3 Goods Receipt	
Management of delivery notifications	☒
Goods receipt planning	☒
Goods receipt acquisition	☒ Acquisition without reference to purchase order ☒ Acquisition with reference to purchase order and display of the outstanding purchase order items/quantities
Flag complete/partial delivery	☒
Goods receipt procedure	☒ Single stage ☒ Two stage with stock in quality inspection
Procedures without stocking	☒ Backorder break-down in the goods receipt ☒ Cross-docking ☒ Flow-through
5.4 Invoice Verification	
Invoices with reference (i.e., with suggested values)	☒ Purchase order ☒ Delivery
Collective invoices with reference (i.e., with suggested values)	☒ Multiple purchase orders ☒ Multiple deliveries
Advance invoice acquisition	☒
Automatic invoice verification	☒ Verification at the total and invoice item level ☒ Online verification in dialog ☒ Batch verification
Allowance for tolerances	☒ Absolute values ☒ Percentage values
5.5 Warehousing	
Article can be stored in several locations	☒
Stockholding strategies	☒ Chaotic stockholding ☒ Fixed location
Differentiated stockholding for a specific article (e.g., picking area with fixed location and reserve area with chaotic stockholding)	☒
Stock types	☒ Freely-usable stock ☒ Blocked stock ☒ Sales order stock ☒ Vendor consignment store ☒ Stock for transfer

Inventory management	☒ Quantity-based ☒ Value-based
Inventory valuation	☒ Standard price ☒ Moving average ☒ FiFo (first-in, first-out) ☒ LiFo (last-in, last-out)
Promotion stocks with their own valuation	☒
Possibility to have negative stock levels	☒
Storage bin properties	☒ Maximum volume ☒ Maximum weight ☒ Suitability for dangerous goods ☒ Automatic consideration of the properties during the stock placement
5.5.1 Returnable Transport Packaging	
Handling of returnable transport packaging / returnable packaging	☒ Summary ☒ Individual postings
Number of returnable packaging types	unlimited
Different deposit rates per returnable packaging type	☒
5.5.2 Physical Inventory	
Physical inventory procedures	☒ Periodic inventory ☒ Permanent inventory ☒ Zero inventory
Physical inventory aids	☒ Print-out of count lists ☒ Support for scanner inventory ☒ Block articles to be counted
Physical inventory history	☒
5.6 Marketing	
5.6.1 Customer	
Customer number	Alphanumeric
Customer categories	☒ Potential customer ☒ Consumer (with limited customer master data) ☒ Customer

Customer addresses/roles	☒ Sold-to party ☒ Service/product recipient ☒ Invoice recipient ☒ Bonus recipient
Number of delivery addresses per customer	arbitrary
Customer characteristics	☒ Industry sector ☒ Region ☒ Size ☒ Customer assortment ☒ Association assignment ☒ other freely-assignable fields
5.6.2 Article Listing	
Article listing	Assignment of articles, article groups or assortments to specific customers/stores or to all customers/stores. Maintenance programs are available to make a listing for customer/store groups.
Listing methods	☒ Positive listing ☒ Negative listing
Article listing with time control	☒ for one-off seasonal articles ☒ for cyclical seasonal articles
5.6.3 Selling Prices / Sales Calculation	
Calculation methods	☒ Net prices + supplement calculation ☒ Optical rounding (price points)
Sales price scales	☒ Quantity scales ☒ Price scales
Level of the sales price definition	☒ Customer ☒ Customer group ☒ All customers
Time control of the sales conditions	☒ Prices ☒ Discounts
5.6.4 Subsequent Renumerations (Sales Bonus Agreements)	
see 5.1.3	
5.7 Sales	
5.7.1 Quotation Management	
Price information management	☒ with display for new request ☒ with display for order entry

5.7.2 Order Entry	
Order types	☒ Immediate order ☒ Standard order ☒ Cash sale ☒ Commission order ☒ Third-party order
Third-party order with reference between	☒ sales order and vendor purchase order ☒ customer invoice and vendor invoice
Special telephone sales support with	☒ compact customer information ☒ CTI connection
5.8 Goods Issue	
5.8.1 Route Planning	
Parameters for the route planning	☒ Volume ☒ Weight ☒ Number and type of the transport containers ☒ Delivery sequences ☒ Time window ☒ other
Interface to external route planning systems	☒
5.8.2 Picking	
Picking document	☒ with picking note (per item) ☒ with picking document (per storage area) ☒ with delivery note (for the complete customer delivery)
5.9 Invoicing	
Delivery assessment	☒ according to order date ☒ according to delivery date
Invoicing	☒ immediate ☒ in invoicing cycles
Collective invoices	☒
Split invoice	☒
5.10 Store Management	
Credit card handling	☒ with credit card authorization (e.g., VISA) ☒ as outstanding financial accounting items (e.g., customer cards)

Functions provided by the POS interface	☒ Download master data ☒ Upload summarized sales activities ☒ Upload non-aggregate sales activities ☒ Upload goods movements ☒ Upload money transactions
Procedures for the store purchasing	☒ Decentral by stores ☒ Central purchasing ☒ Central consolidation of the decentral individual purchases
Allocation table for central purchasing	☒ Manual allocation ☒ Automatic with fixed allocation quantities ☒ Automatic with percentage allocation quantities
Inventory management in the store	☒ Quantity-based ☒ Value based on the purchase price ☒ Value based on the selling price
Stock transfer to the store as...	☒ stock movement, automatic transfer posting (single-stage stock transfer) ☒ stock movement, manual stock transfer goods issue + goods receipt (two-stage stock transfer; not relevant for the financial accounting) ☒ sales (two-stage stock transfer; relevant for the financial accounting)

Terms and Abbreviations

A-Gate	Application Gateway
ABAP	Advanced Business Application Programming
ALE	Application Link Enabling
ANSI	American National Standards Institute
ARIS	Architecture of Integrated Information Systems
ARPANET	Advanced Research Projects Agency-NET
ATP	Vailable-to-Promise
BAPI	Business Application Programming Interface
BDC	Batch Data Communication
BDS	Business Document Service
BIC	Business Information Collection
BO	Business Object
BW	Business Information Warehouse
C&C	Cash and carry
CCG	Centrale for Coorganisation
CIC	Customer Interaction Center
CM	Category Management
CMW	Category Manager Workbench
COM/DCOM	Component Object Model / Distributed Component Object Model
CORBA	Common Object Request Broker Architecture
CTI	Computer Telephony Integration
DIAG	Dynamic Information and Action Gateway
DP	Data processing
DPP	Direct Product Profit
DRS	Distributed retail information system
EA	Efficient Assortment
EAN	European Article Number
EC	Electronic Commerce
ECR	Efficient Consumer Response
EDI	Electronic Data Interchange
EDIFACT	Electronic Data Interchange For Administration, Commerce and Transport

EIS	Executive Information Systems
EP	Efficient Promotion
EPC	Event-Controlled Process Chain
EPI	Efficient Product Introduction
ER	Efficient Replenishment
ERM	Entity-Relationship Model
ERP	Enterprise resource planning
EU	European Union
EWB	Editorial Workbench
EWS	Early-warning system
FiFo	First in - First out
FIS	Financial Information System
FTAM	File Transfer Access and Management
FTP	File Transfer Protocol
GfK	Gesellschaft für Konsumforschung (German market research company)
HTML	HyperText Markup Language
HTTP	HyperText Transfer Protocol
HR	Human resources
IAC	Internet Application Component
IDoc	Intermediate Document
ILN	International Location Number
IP	Internet Protocol
IRB	Information Requirements Builder
ISAPI	Internet Server Application Programming Interface
ISDN	Integrated Services Digital Network
ISO	International Standardization Organization
IT	Information Technology
ITS	Internet Transaction Server
LAN	Local area network
LiFo	First in - Last out
LIL	Logistics Information Library
LIS	Logistics Information System
MDA	Mobile Data Acquisition
MIME	Multipurpose Internet Mail Extensions
MRP	Material Resource Planning
MSS	Management Support System
NSAPI	Netscape Server Application Programming Interface
OIW	Open Information Warehouse
OLAP	Online Analytical Processing
OLE	Object Linking and Embedding
OTB	Open-to-Buy
PC	Personal Computer
PIS	Personnel Information System
POI	Point of information
POS	Point of sale

PS	Project System
RFC	Remote Function Call
RIS	Retail Information System
ROI	Return-On-Investment
S/MIME	Secure/Mulripurpose Internet Mail Extensions
SD	Sales and Distribution
SEDAS	Standard regulations for the standardized data exchange systems
SEM	Strategic Enterprise Management
SET	Secure Electronic Transaction
SINFOS	Master data information system of the CCG
SMTP	Simple Mail Transfer Protocol
SNP	Supply Network Planning and Deployment
SPB	Source Profile Builder
SQL	Structured Query Language
SSL	Secure Socket Layer
TCP/IP	Transmission Control Protocol / Internet Protocol
URL	Uniform Resource Locator
VMI	Vendor Managed Inventory
WAN	Wide area network
WWW	World Wide Web
XML	eXtensible Markup Language

Bibliography

Abreo, E. (2001) : TheStandard.com. Bibliofind Closes Its Books After Hack. http://biz.yahoo.com/st/010306/22672.html, downloaded on 2001-02-05.

Ahlert, D. et al. (1998): Informationssysteme für das Handelsmanagement. Springer, Berlin et al.

Ahlert, D. (1997): Warenwirtschaftsmanagement und Controlling in der Konsumgüterdistribution. In: Ahlert, D.; Olbricht, R. (publisher.) Integrierte Warenwirtschaftssysteme und Handelscontrolling. 3rd Edition. Schäffer-Poeschel, Stuttgart, pp. 3-113.

Angeli, A.; Streit, U. (2001): The SAP R/3 guide to EDI and interfaces - cut your implementation cost with IDocs, ALE and RFC. 2. rev. ed. Vieweg, Braunschweig/Wies baden.

Armstrong, A.; Hagel, J. (1995): Real Profits from Virtual Communities. In: The McKinsey Quarterly, Volume 3, pp. 127-141.

AWV (1996): Warenwirtschaftssystem-gestützte Inventur im Handel. Eschborn

Barth, K. (1996): Betriebswirtschaftslehre des Handels. 3rd Edition. Gabler, Wiesbaden.

Becker, J.; Schütte, R. (1996): Handelsinformationssysteme. Moderne Industrie, Landsberg/ Lech.

Bryan, M. (1997): Guidelines for using XML for Electronic Data Interchange, XML/EDI Group, 1997-12-23. http://www.xmledi.net, downloaded on 1998-01-11.

Buxmann, P.; Ladner, F.; Kronenberg, R.; Weitzel, T. (1999): Extensible Data Interchange – Die Zukunft von EDI. Management Workshop on XML, Frankfurt am Main, 1999-01-15.

CCG (1995): Efficient Consumer Response (ECR) – eine Materialsammlung. Centrale für Coorganisation GmbH (publisher), Cologne, pp. 79-91.

CCG (1997): Standard-Warengruppenklassifikation für Verbrauchsgüter, Gebrauchsgüter, Investitionsgüter und Rohstoffe. 2nd Edition. Centrale für Coorganisation GmbH, Cologne.

CCG (1999): EDI. http://www.ccg.de/english/index_e.htm, downloaded on 2001-02-03.

Chapman, B.; Zwicky, E.; Russel, D. (1995): Building Internet Firewalls. O'Reilly & Associates, Sebastopol, CA.

Chen, P.P.-S. (1976): The Entity Relationship Model – Toward a Unified View of Data. In: ACM Transactions on Database-Systems, 1st Issue, Volume 1, pp. 9-36.

Collaborative Planning, Forecasting and Replenishment (CPFR) Committee (1999): Our Mission; http://www.cpfr.org/, downloaded on 2001-03-20.

Diamond, J.; Pintel, G. (1996): Retailing. 6th Edition. Prentice Hall, Upper Saddle River, NJ.

Deutsch, M. (1995): Unternehmenserfolg mit EDI: Strategie und Realisierung des elektronischen Datenaustausches. Vieweg, Berlin.

Dräger, E. (1998): Projektmanagement mit SAP R/3. Addison-Wesley, Bonn.

ECR Europe (1999): What is ECR?. http://www.ecrnet.org/ECR/ECR.home, downloaded on 2001-03-20.

eMarketer (1999): eCommerce: Top Spending Categories. http://www.emarketer.com/estats/ec_proj.html, downloaded on 1999-09-16.

EU (2000): 1999/93/EG regulation of the European Parliament and the Council from 13.12.1999 about the communities general conditions for electronic signatures. In: Official gazette of the European Union from 19.01.2000.

Falk, B.; Wolf, J. (1992): Handelsbetriebslehre. 11th Edition. Moderne Industrie, Landsberg/ Lech.

Fischer, T. (1997): Computergestützte Warenkorbanalyse als Informationsquelle des Handels-management. In: Ahlert, D.; Olbrich, R. (publisher) Integrierte Warenwirtschafts-systeme und Handelscontrolling. 3rd Edition. Schäffer-Poeschel, Stuttgart, pp. 282–312.

Frackmann, E. (1996): Managementcomputing. De Gruyter, Berlin et al.

Günther, T. (1993): Direkter Produkt-Profit. In: zfbf, 45th Issue, Volume 5, pp. 460-482.

Hahne, M. (1998): Logische Datenmodellierung für das Data Warehouse. Bestandteile und Varianten des Star Schemas. In: Chamoni, P.; Gluchowski, P. (publisher) Analytische Informationssysteme. Springer, Berlin et al.

Hadjiconstantinou, E. (1999): Quick Response in the Supply Chain. Springer, Berlin et al.

Harting, D. (1989): Lieferanten-Wertanalyse. Schäffer, Stuttgart.

heise online news (1999): Datenschutzprobleme bei Yello Strom. http://www.heise.de/newsticker/data/nl-20.09.99-000/, downloaded on 2001-02-20.

Hernandez, J. (2000): The SAP R/3 Handbook - Comprehensive and Practical Guide to Basis System Administration for SAP R/3. Covers Versions 4.0/4.5/4.6, ABAP/4, Workbench Organizer and CTS. 2. ed., McGraw-Hill, Frankfurt.

Hernandez, J. (1998): SAP R/3 3.1 für Administratoren. Sybex, Dusseldorf.

Hertel, J. (1999): Warenwirtschaftssysteme. 3rd Edition. Physica, Heidelberg.

Heydt, A. von der (1998): Efficient Consumer Response (ECR). 3rd Edition. Peter Lang Verlag, Frankfurt/M.

Immon, W. H. (1992): Building the Data Warehouse. John Wiley & Sons, New York et al.

Kasturi, R. (1999): SAP R/3 ALE & EDI Technologies. Mc Graw-Hill, Frankfurt.

Kretschmer, R. (1996): Developing SAP's R/3 applications with ABAP/4. Sybex, Düsseldorf.

Lilien, G. L.; Kotler, P.; Moorthy K. S. (1992): Marketing Models. Prentice-Hall, Englewood Cliffs, NJ.

Mason, J. B.; Burns, D. J. (1998): Retailing, 6th Edition, Dame Publications, Houston.

Meier, M.; Fülleborn, A. (1999): Integration externer Führungsinformationen aus dem Internet in SAP Strategic Enterprise Management (SEM). In: Wirtschaftsinformatik, 41th Issue, Volume 5, pp. 449-457.

Meffert, H. (1998): Marketing. Grundlagen der Absatzpolitik. 8th Edition. Gabler, Wiesbaden.

Mende, U. (2000): Software Development for SAP R/3 - Data Dictionary, ABAP/4, Interfaces. Springer Berlin.

Merrilees, B.; Miller, D. (1996): Retailing Management. A Best Practice Approach. RMIT Press, Victoria.

Mertens, P. (1999): Integration interner, externer, qualitativer und quantitativer Daten auf dem Weg zum Aktiven MIS. In: Wirtschaftsinformatik, 41th Issue, Volume 5, pp. 405-415.

Mertens, P. et al. (1998): Grundzüge der Wirtschaftsinformatik. 5th Edition. Springer, Berlin et al.

Meyer, B. (1997): Object-oriented Software Construction. 2nd Edition, Prentice Hall London.

Milde, H. (1996): Kategorie-Management in der Praxis. In: M&M (Marktforschung und Management), 40th Issue, Volume 1, pp. 10-17.

Möhlenbruch, D. (1999): Kundenorientierung durch Category Management – Kritische Analyse eines Kooperationsmodells zwischen Industrie und Handel. In: Trommsdorf, V. (Hrsg.) Handelsforschung 1997/98: Kundenorientierung im Handel. Gabler, Wiesbaden, pp. 113-133.

Möhlenbruch, D.; Kotschi, B. (1999): Die Verkaufsförderung als Kooperationsbereich zwischen Industrie und Handel. In: Trommsdorf, V. (publisher) Handelsforschung 1999/00: Verhalten im Handel und gegenüber dem Handel. Gabler, Wiesbaden.

Moser, G. (1999): SAP R/3 Interfacing using BAPIs. Vieweg, Braunschweig/Wiesbaden.

Mucksch, H.; Behme, W. (publisher) (1997): Das Data Warehouse-Konzept. 2nd Edition. Gabler, Wiesbaden.

Mulcahy, D. E. (1993): Warehouse Distribution and Operations Handbook. McGraw-Hill, New York.

Nenninger, M.; Gerst, M. (1999): Wettbewerbsvorteile durch Electronic Procurement – Strategien, Konzeption und Realisierung. In: Hermanns, A.; Sauter, M. (publisher) Management-Handbuch Electronic Commerce. Vahlen, Munich.

Nusser, S. (1998): Sicherheitskonzepte im WWW. Springer, Berlin et al.

Perez, M. and others (1999): SAP R/3 system on the Internet. Addison-Wesley Longman, Bonn.

Perridon, L.; Steiner, M. (1995): Finanzwirtschaft in der Unternehmung. 8th Edition. Vahlen, Munich.

Pretzel, J. (1996): Gestaltung der Hersteller-Handel-Beziehung durch Category Management. In: Markenartikel, 58th Issue, Volume 1, pp. 21-25.

Recht, P.; Zeisel, S. (1997): Warenkorbanalysen in Handelsunternehmen mit dem Conjoint Profit Modell. In: Controlling, 9th Issue, Volume 2, pp. 94-100.

SAP AG (1998): System R/3 – SAP Retail. Funktionen im Detail. SAP AG, Walldorf.

SAP AG (2000a): Waren- und Sortimentsplanung. Release 4.6C. SAP AG, Walldorf.

SAP AG (2000b): SAP Business Information Warehouse. http://www.sap-ag.de/germany/products/bw/index.htm, downloaded on 2000-01-12.

Scheer, A.-W. (2000): ARIS - Business Process Modeling. 3rd Edition. Springer, Berlin et al.

Scheer, A.-W. (1999): ARIS - Business Process Frameworks. 3rd Edition. Springer, Berlin et al.

Schütte, R.; Vering, O.; Wiese, J. (2000): Erfolgreiche Geschäftsprozesse durch standardisierte Warenwirtschaftssysteme. Marktanalyse – Produktübersicht – Auswahlprozess. Springer, Berlin et al.

Shih, C.; Jansson, M.; Drummond, R. (1997): Mime-based Secure EDI. EDIINT Working Group, draft-ietf-ediint-as1-05.txt, 1997-07-08. ftp://ftp.ietf.org/internet-drafts/draft-ietf-ediint-as1-05.txt , downloaded on 1998-01-16.

Steffenhagen, H. (1995): Konditionengestaltung zwischen Industrie und Handel. Ueberreuter Wirt., Vienna.

Treis, B.; Holzkämper, O. (1998): Strategische Überlegungen des Handels zum Category Management – Das Aktivitätenportfolio. In: Trommsdorf, V. (publisher) Handelsforschung 1998/99: Innovation im Handel. Gabler, Wiesbaden, pp. 267-289.

Wenzel, P. (publisher) (1999): Business Computing mit SAP R/3. Vieweg, Braunschweig/ Wiesbaden.

Wolf, T.; Hofmann, R.; Röder, H. (1997): Electronic Commerce – Status quo und Perspektiven '97. KPMG Consulting GmbH, http://www.kpmg.de , downloaded on 2001-02-14.

Index

C

D

E

M

N

O

P

Q

R

S

SAP® EXCELLENCE

H.-J. Appelrath, Institut OFFIS, Oldenburg, Germany; **J. Ritter,** OSC GmbH, Oldenburg, Germany

SAP R/3 Implementation

Methods and Tools

2000. XII, 188 pp. 48 figs., 5 tabs. (SAP Excellence)
Hardcover * **DM 79,-**
ISBN 3-540-66863-2

G. Knolmayer, University of Bern, Switzerland;
P. Mertens, A. Zeier, University of Erlangen-Nürnberg, Nuremberg, Germany

Supply Chain Management Based on SAP Systems

Order Processing in Manufacturing Companies

2000. Approx. 200 pp. (SAP Excellence) Hardcover
* **DM 79,-**
ISBN 3-540-66952-3

P. Buxmann, W. König, University of Frankfurt, Germany

Inter-organizational Cooperation with SAP Systems

Perspectives on Logistics and Service Management

2000. XII, 186 pp. 92 figs., 1 tab. (SAP Excellence)
Hardcover * **DM 79,-**
ISBN 3-540-66983-3

J. Becker, University of Muenster, Germany;
W. Uhr, University of Dresden, Germany;
O. Vering, University of Muenster, Germany

Retail Information Systems Based on SAP Products

2000. Approx. 250 pp. (SAP Excellence) Hardcover
* **DM 79,-**
ISBN 3-540-67199-4

Please order from
Springer · Customer Service
Haberstr. 7
69126 Heidelberg, Germany
Tel: +49 (0) 6221 - 345 - 217/8
Fax: +49 (0) 6221 - 345 - 229
e-mail: orders@springer.de
or through your bookseller

Springer

* Recommended retail prices. Prices and other details are subject to change without notice. In EU countries the local VAT is effective. d&p · BA 67189/1